"Amid all the apocalyptic visions about tomorrow's workplace—from widespread unemployment to pitched battles between man and machine—comes a bracing dose of common sense. *Reinventing Jobs* cuts through the hype and hysteria and shows leaders how to apply automation and artificial intelligence in their organizations. Jesuthasan and Boudreau are the clear-eyed guides every organization needs to track their way through the future of work."

—DANIEL H. PINK, author, *Drive* and *When*

"In the age of the Internet of Things (IoT) and AI, the traditional notions of 'Economic Man' and 'Social Man' need to be superseded by the concept of 'Self-motivated' Man. *Reinventing Jobs* is a book ahead of its time and points the way to the 'humanization' of the enterprise."

—ZHANG RUIMIN, Chairman and CEO, Haier Group

"*Reinventing Jobs* provides a step-by-step guide to rethinking how organizations deal with automation. Jesuthasan and Boudreau provide leaders with a thoughtful framework that helps us ask the right questions when we're considering where and when to automate. They recognize that automation is not a simple 'rip and replace' of certain jobs and tasks, but rather something that requires a layered approach."

—JIM WHITEHURST, President and CEO, Red Hat

"This book provides an excellent, much-needed framework to guide the collaboration necessary among CIOs, CHROs, and other functional leaders to optimize work and the application of automation in today's rapidly changing environment."

—RALPH W. BROWN, Chief Technology Officer and Senior Vice President, Research & Development, CableLabs

"A must-read for every CEO. A most thoughtful approach with examples from numerous industries that illustrate that automation and workflow optimization are critical to competing effectively."

—**ANTHONY G. PETRELLO,** Chairman, President, and CEO, Nabors Industries

"*Reinventing Jobs* is a book that CIOs, CHROs, and all C-suite leaders should read together. This four-step framework takes the guesswork out of building your digital strategy and a future-ready, engaged, and agile workforce."

—**DIANA MCKENZIE,** Senior Vice President and CIO, Workday

"This book provides a thoughtful, step-by-step guide to responsible automation that is relevant to all leaders in government, academia, and the private sector. Jesuthasan and Boudreau bring an insightful and practical approach to dealing with one of the most significant challenges of the Fourth Industrial Revolution."

—**SAADIA ZAHIDI,** Head of Education, Gender and Work and member of the Executive Committee, World Economic Forum

"We live in a world where advanced technologies are disrupting entire industries, yet few have advanced the discussion beyond a surface consideration of technology either eliminating or creating jobs. Now, Jesuthasan and Boudreau offer a comprehensive approach to transforming work while enabling organizations—and individuals—to adapt and thrive."

—**ALAN MAY,** Chief People Officer, Hewlett Packard Enterprise

Reinventing Jobs

Reinventing

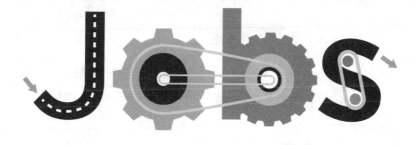

A 4-STEP APPROACH FOR APPLYING AUTOMATION TO WORK

Ravin Jesuthasan and John W. Boudreau

HARVARD BUSINESS REVIEW PRESS

Boston, Massachusetts

Library of Congress Cataloging-in-Publication Data

Names: Jesuthasan, Ravin, 1968– author. | Boudreau, John W., author.
Title: Reinventing jobs : a 4-step approach for applying automation to work /
    by Ravin Jesuthasan and John W. Boudreau.
Description: Boston, Massachusetts : Harvard Business Review Press, [2018]
Identifiers: LCCN 2018010210 | ISBN 9781633694071 (hardcover : alk. paper)
Subjects: LCSH: Industrial management—Technological innovations. |
    Automation. | Artificial intelligence. | Technological innovations—Employee
    participation.
Classification: LCC HD45 .J47 2018 | DDC 658.5/14—dc23
LC record available at https://lccn.loc.gov/2018010210

ISBN:9781633694071
eISBN: 9781633694088

To my family, friends, colleagues, and church for
your inspiration, love, and support.
—*Ravin Jesuthasan*

To my family and the students and colleagues
who honor me every day by sharing their
wisdom and support. It is a gift to share our
collective journey of reinvention.
—*John Boudreau*

*Once the bedrock of competitive advantage,
legacy, whether mindset or infrastructure, is
increasingly the primary obstacle to sustainable
automation and the future of work.*

CONTENTS

- - - - - - - - - - - - - -

Acknowledgments   *xi*

INTRODUCTION
AI and Robotics Are Here. Now What?   1

PART ONE
Optimizing Work Automation
A Four-Step Framework

ONE
Deconstruct the Job   19
Which Job Tasks Are Best Suited to Automation?

TWO
Assess the Relationship between
Job Performance and Strategic Value   41
What Is the Automation Payoff?

THREE
Identify Options   57
What Automation Is Possible?

FOUR
Optimize Work   87
What Does the Right Human-Automation
Combination Look Like?

Contents

PART TWO

Redefining the Organization, Leadership, and Workers

Automation Implications beyond
Reinventing Jobs

FIVE

The New Organization 123

Digital, Agile, Boundaryless, and Work-Centric

SIX

The New Leadership 141

Democratic, Social, and Perpetually Upgraded

SEVEN

Deconstruct and Reconfigure Your Work 167

Using the Work-Automation Framework to
Navigate Your Personal Work Evolution

Appendix 187

Notes 193

Index 201

About the Authors 211

ACKNOWLEDGMENTS

We are grateful for the support and encouragement of our colleagues at Willis Towers Watson. We particularly appreciate the assistance provided by Anne-Marie Jentsch, George Zarkadakis, Juliet Taylor, Tracey Malcolm, Kannie Kong, Maggy Fang, and Edward Liu in the writing of this book. Thanks as well to David Creelman for his help, encouragement, and insights.

Thanks to the team at Harvard Business Review Press, particularly our wonderful, tireless, supportive, and insightful editor Melinda Merino, who saw the potential of this book far earlier and more clearly than anyone.

AI and Robotics Are Here. Now What?

If you are a leader wrestling with where, when, and how to apply automation in your organization, you're in good company. Leaders everywhere are asking how automation will affect their organizations and how jobs—those of their teams, bosses, colleagues, friends, and families as well as their own—might change or even be eliminated. Optimists say that machines will free human workers to do higher-value, more creative work. Pessimists predict massive unemployment or even an apocalypse in which humans merely serve the robots. Of course, both the optimists and the pessimists are partially correct and partially wrong.

But what everyone gets wrong is asking, "In which jobs will automation replace humans?" We see smart and well-meaning leaders get stuck in the typical discourse about job replacement. It's a dead-end conversation. Simply asking which humans will be replaced fails to take into account how work and automation will evolve. You can't solve the "how to automate work" problem by thinking only about automation replacing jobs.

Consider the example of the automated teller machine (ATM). A familiar example, for sure, but one that illustrates the myopia that comes from asking, "Which jobs will be replaced by automation?" It's also a good example to start with because the evolution of banking work, through automation, is continuous.

Did the ATM Mean the End of Bank Tellers?

On June 14, 2011, Barack Obama noted that ATMs allowed businesses to "become much more efficient with a lot fewer workers."[1] In reality, for decades, the number of teller jobs actually *increased*, along with the number of ATMs. In 1985, the United States had 60,000 ATMs and 485,000 bank tellers. In 2002, the United States had 352,000 ATMs and 527,000 bank tellers.[2] Understanding how automation affects work clearly requires a more nuanced approach than "how many jobs do ATMs replace?"

Economist James Bessen explains the paradox of more ATMs creating more teller jobs in his book *Learning by Doing*.[3] Quoted in a *Wall Street Journal* article, he says, "The average bank branch used to employ 20 workers. The spread of ATMs reduced the number to about 13, making it cheaper for banks to open branches. Meanwhile, thanks in part to the convenience of the new machines, the number of banking transactions soared, and banks began to compete by promising better customer service: more bank employees, at more branches, handling more complex tasks than tellers in the past."[4]

Fast-forward to today. Personal devices and cloud-based financial transactions demand even greater changes in banking work. Has automation finally replaced tellers? Again, the reality is more nuanced. In May 2017, "while more than 8,000 U.S. bank branches have closed over the past decade (an average of more than 150 per state) and more than 90% of transactions now take place online, the number of U.S. bank employees has remained relatively stable at more than 2 million."[5]

Why is there such stability in the number of bank employees as automation advances? The work of the teller job has evolved. "Where the bank branches still stand as a brick-and-mortar presence, the tellers have started coming out from behind the window with smartphone or tablet in hand to help customers help themselves. But with thousands of those branches closing, you're more likely to find a teller online now. They've become the human face of an increasingly virtual world. It's a role exemplified in Bank of America's new experiment with hybrid banking, small unstaffed mini-branches that offer a direct link to tellers via video conference."[6]

The ATM story is a useful parable for leaders, workers, and everyone else, because it illustrates why the simplistic idea of "technology replaces human worker jobs" is so misleading. That approach can't predict how work and automation actually evolve. The story also illustrates the pivotal future capability for leaders—optimizing the ever-evolving options for combining human and automated work.

That's what this book is about. Going beyond the question of how and to what extent automation will replace traditional jobs, we present a systematic framework in the form of a structured, four-step approach that leaders can use to reveal optimal work-automation combinations and redefine jobs in their organizations. Our framework frees you from asking the simplistic question of "which jobs will be replaced by automation?" and instead gives you a more nuanced, but precise and actionable method for determining the optimal combination of humans and machines in your organization.

The Road Map for Reinventing Jobs

This book is for everyone who must consider how automation will affect jobs and work, which includes almost anyone. However, we wrote this book particularly for leaders, because leaders must decide where, why, and how to *optimize the combination of human and automated work.* This can appear messy at first. Again, it is not as simple as asking which jobs will technology replace. Nor is it as simple as "lift and shift" outsourcing, where intact jobs move to third parties. Instead, automating work requires leaders to radically but systematically rethink the "job," the main unit of work for centuries. Leaders that understand this and take a disciplined, nuanced approach will reap enormous rewards.

We know this because we have spent decades helping business leaders achieve strategic success through people and work. Ravin has been recognized as one of the twenty-five most influential management consultants in the world and has helped some of the largest and most prominent companies worldwide transform their organizations and work to realize a step change in performance. He has worked with governments, educational institutions, and nongovernmental organizations like the World Economic Forum on the future of work. John is one of the world's foremost thought leaders on strategic human capital, work, and the future of the human resources profession. He has helped clarify how work, talent, and the organization are most pivotal to strategic success in many companies, ranging from early-stage startups to some of the largest global organizations.

Our focus on the impact of work, talent, and the organization on strategic success brings a truly unique and differentiated perspective to the question of how to automate work. Most experts approach this problem from the technology side. We come at it from the organizational and human capital perspective. We have a strong point of view about how to achieve optimal work-automation combinations based on our decades of experience helping organizations reinvent work, leadership, and even themselves in light of the latest workplace technologies and innovations.

While the specific challenges facing the organizations that we've worked with have changed over the years— whether in response to process improvement (business process reengineering or agile), redesign, or new methods of resourcing (like outsourcing, talent platforms, or contingent workers)—the approach that we've developed and used over the past ten years to help leaders reinvent their companies is a structured, four-step framework: (1) deconstruct jobs into component work tasks; (2) assess the relationship between job performance and strategic value; (3) identify options for recombining tasks in light of the new technology or process; and (4) optimize work by putting it all together to reinvent jobs.

Most recently, we've used this systematic step-by-step approach to help leaders with the specific challenge of how

Is Your Organization Ready for Automation?

In addition to helping organizations grapple with the future of work, this book also focuses on research from Willis Towers Watson.

One recent study highlights the preparedness of organizations for the impact of several trends, including automation. The figure from our Global Future of Work Survey (page 7) shows how the participating companies rated their preparedness in several key areas: already fully prepared, already taking some actions, planning to take actions, considering taking actions, or unprepared. The two areas where companies were most unprepared are "deconstructing jobs and identifying which tasks can best be performed by automation" and "identifying reskilling pathways for talent whose work is being transformed by automation." As we will show in this book, these two areas underpin future leadership requirements as jobs are reinvented; they are pivotal to our framework.

to respond proactively to the onset of work automation. We've used this framework in organizations across a variety of industries—bio pharmaceuticals, oil and gas, high tech, financial services, transportation—to optimize the power and potential of automation and to solve the work and challenges associated with it. (See the sidebar "Is Your Organization Ready for Automation?" for more about our research.)

Now, with this book, we offer you a hands-on guide to the four-step approach and show you how to apply it to your automation challenge. Our hope is that by using this book, you can deconstruct the work in your organization, identify

Actions taken and opportunity areas related to creating optimal combinations of humans and automation

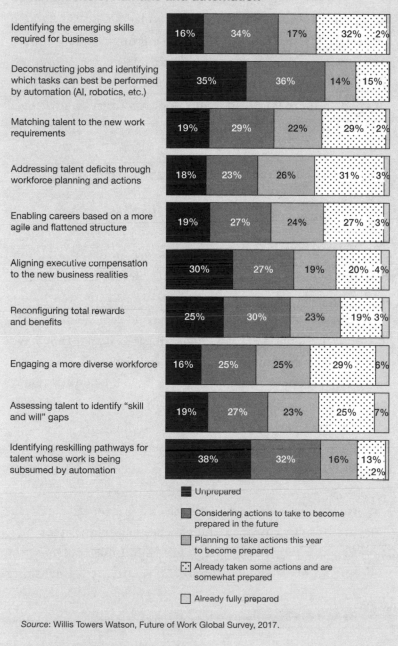

Identifying the emerging skills required for business	16%	34%	17%	32%	2%
Deconstructing jobs and identifying which tasks can best be performed by automation (AI, robotics, etc.)	35%	36%	14%	15%
Matching talent to the new work requirements	19%	29%	22%	29%	2%
Addressing talent deficits through workforce planning and actions	18%	23%	26%	31%	3%
Enabling careers based on a more agile and flattened structure	19%	27%	24%	27%	3%
Aligning executive compensation to the new business realities	30%	27%	19%	20%	4%
Reconfiguring total rewards and benefits	25%	30%	23%	19%	3%
Engaging a more diverse workforce	16%	25%	25%	29%	6%
Assessing talent to identify "skill and will" gaps	19%	27%	23%	25%	7%
Identifying reskilling pathways for talent whose work is being subsumed by automation	38%	32%	16%	13%	2%

- Unprepared
- Considering actions to take to become prepared in the future
- Planning to take actions this year to become prepared
- Already taken some actions and are somewhat prepared
- Already fully prepared

Source: Willis Towers Watson, Future of Work Global Survey, 2017.

7

the payoffs, choose the right automation approaches, and then optimize work automation. Our framework will help you navigate the ever-changing, complex, and nuanced opportunities of automation. It provides a tool set to help you resist the siren's song of simply cutting costs by substituting automation for humans that is still a common pitch of many a robotic process automation (RPA) vendor. Such simplistic approaches often produce unforeseen problems that can be anticipated with the sophisticated approach we describe.

How This Book Unfolds

Part 1 presents the four-step framework in detail and how to apply it to your automation-work choices. The first step is to deconstruct jobs into component tasks. This step recognizes that "in which jobs will automation replace humans?" is the wrong question. Jobs contain many tasks that have different automation compatibility and payoff. You can see these patterns only when you deconstruct jobs into tasks. The better question, therefore, is which job tasks are best suited to automation?

We can then describe tasks in terms of their automation compatibility using these three dimensions:

- Repetitive–variable. Is the task more repetitive, with predictable routines and success criteria, or more variable, with unique and unpredictable routines and changing success criteria, requiring innovation and perhaps the application of decision rules to new or unique circumstances?

- Independent-interactive. Is the work task performed more independently by a single person or more interactively with others, involving communication and empathy?

- Physical-mental. Is the work task more physical, using strength and manual dexterity, or more mental, using cognition, creativity, and judgment?

Chapter 1, "Deconstruct the Job," discusses how to deconstruct jobs into work tasks and then how to categorize them using these three questions, the foundation for optimally applying automation.

The next step, in chapter 2 ("Assess the Relationship between Job Performance and Strategic Value") is to answer the "what's the payoff?" question. If your goal is to prevent mistakes, that implies a very different work-automation approach than if the goal is to improve existing performance, and both are very different from augmenting your human workers to achieve exponential performance value. This chapter describes how to analyze each task to identify its relationship between work performance and value. That clarifies your goals and the optimum payoff to work automation.

The third step of our framework asks, "What automation is possible?" Most leaders jump directly to this step—imagining all sorts of automation options. Yet, you can only optimize automation options after you've done the first two steps. Once you've deconstructed jobs into work tasks, identified how compatible the tasks are with automation, and identified the performance payoff of those tasks, you can more precisely identify automation options. We'll show you three types of automation: robotic process automation, cognitive automation, and social robotics.

RPA applies process automation to high-volume, low-complexity, and repetitive tasks. Cognitive automation takes on more complex tasks by applying intelligence like pattern recognition or language understanding. Social robotics involves robots interacting or collaborating with humans in physical space by combining sensors, artificial intelligence (AI), and mechanical mobile robots.

Chapter 3, "Identify Options," will help you understand these three automation types and their relevance to different types of work. You'll see how the information from steps one, two, and three gives clues about whether automation should replace, augment, or create new work for humans.

Should automation substitute human endeavor, augment it, or create new human work? What does work optimization look like? The fourth step of the framework pulls it all together. Chapter 4, "Optimize Work," presents actual examples that illustrate optimizing work automation. It shows how the four elements of our framework combine to help you see work automation as more than just a jumble of anecdotes and examples, but rather discern how each example reflects an optimum combination of job deconstruction, return on performance, and automation.

However, optimizing work automation doesn't end with reinventing the job. Reinventing jobs redefines the very nature of your organization. The new human-automation work that you create seldom fits easily into traditional job descriptions and organization structures, and it is often optimally sourced in different ways than traditional employment. Organizations consist of many interconnected jobs and structures. Our experience reveals that true optimization requires connecting the reinvented jobs to structures, decision rights, social networks, culture, and other organization-level factors. It requires fundamental changes in the definition and execution of leadership. Finally, it requires that everyone approach their own work and careers as a constant process of deconstruction and reinvention. Part 2 addresses these issues. Chapter 5, "The New Organization," discusses how optimizing work automation leads to changes in the organization and describes several companies at the forefront of these changes.

These new, reinvented work options can fundamentally shift organizational characteristics such as leadership, power, accountability, culture, structure, information sharing, and decision making. Front-line employees will now have access to information and expertise that previously resided only with their supervisors or managers. Decisions that previously required escalation from the front-line worker to a supervisor to a manager now are assisted by AI. Norms such as "customers come to us because only we have the information they need," must change to "customers arrive with more information than we have, and come to us for a trusted and collaborative relationship."

Chapter 6, "The New Leadership," examines how leaders must evolve to fit this new reality and describes a vital new role for them as guides to perpetually upgrading work. Leaders and workers must collaborate as never before to navigate a world of constantly upgraded work, as the tasks humans formerly did must evolve to be substituted for or augmented by automation.

Workers must trust their leaders and the organization enough to share their ideas about how to automate their own work. Leaders and organizations must devise pathways that provide continuity for workers, but without assuming the work will be constant. Workers must be more entrepreneurial, prepared to shift between employment, contracting, free-lancing, and so on.

Leaders need to rethink their roles and means for realizing the mission of their organizations. This will require new capabilities and tools for leaders and workers, and an increased level of collaboration. Increasingly, both leaders and workers must orchestrate an ecosystem, populated with robots and AI, rather than manage within a self-contained organization.

Chapter 7, "Deconstruct and Reconfigure Your Work," shows how our four-step framework can guide you in thinking about the new meaning of your own work and career and in optimizing your personal strategy for work automation.

We have found that virtually every organization is wrestling and experimenting with automation, but missing the benefits that come from deep and systemic change. One reviewer of our book put it well: "Automation is driven by the strategic need to move faster, be more consumer focused and leverage technology, reduce cost, increase speed, and improve service to create new value in this technology-enabled era. Whatever the goal, it always ultimately rests on the leadership decisions about how automation will affect the work, and the supporting work systems. Yet, very few automation strategies even consider the work, let alone provide a framework for optimizing it."

Lacking a framework and playbook, it's hard to learn from those experiments, particularly the lessons about how, when, and where to apply automation in organizations, and how to create the leadership and organizational structures that will maximize the benefits and reduce the risks. We have seen the value of reframing the work automation challenge through deconstructed and reinvented jobs that use the right automation to optimize and balance performance and risk.

Automation will significantly disrupt and potentially empower the global workforce. It won't happen all at once or in every job, but it will happen. You need a work-automation strategy that recognizes the nuances, realizes the benefits, and avoids needless cost and disruption.

This book will help you build that strategy. The framework we present here will help you better understand the implications of emerging trends, how work automation can transform your organization, and how to drive that transformation.

You can access a digital copy of our framework and numerous other resources to support your journey in reinventing jobs by visiting the following websites: willistowerswatson .com/reinventing-jobs and drjohnboudreau.com/speaking /reinventing-jobs-to-optimize-work-automation.

PART ONE

Optimizing Work Automation

A Four-Step Framework

A t a recent conference on AI and the future of work at the Massachusetts Institute of Technology, experts concluded that technology "both creates and destroys jobs," that labor productivity growth has actually slowed despite technological advances, and that to realize the benefits of automation, we will need to reinvent organizations, institutions, and metrics.[1] Technologies like self-driving cars, chatbots, automated parking lot attendants, and robotic home caregivers are enticing and attention grabbing, but experts say that the pivotal factor in realizing their value is increasingly how leaders optimize the combinations of human and automated work, and then organize and lead to support those combinations.

Part One describes our four-step framework for achieving that optimization. Each chapter describes a vital component of the framework. Then, chapter 4 pulls them all together to show their combined ability to reveal new and more optimal solutions. Every chapter uses actual examples to show how you can better reach your strategic goals by focusing more clearly on work-automation optimization.

To show how the four steps of our framework build on each other, each of the first four chapters begins with a vignette that continues the story of the automatic teller machine (ATM) that we opened with in the introduction. Each chapter relates a new facet of the ATM story and illustrates the ideas that you will see in that particular chapter.

- - - - - - - -

ONE

Deconstruct the Job

Which Job Tasks Are Best Suited to Automation?

Here's a brainteaser: You are given a candle, a box of tacks, and a book of matches. How do you attach the candle to a wall so that you can light it without dripping wax onto the floor below?

The solution to Duncker's candle problem is to deconstruct the box of tacks into its parts (box, tacks).[1] Then you'll see that the tacks can attach one side of the box to the wall and attach the candle to the bottom of the box. In experiments, people who receive the box with the tacks inside solve the problem far less often than those given the box with a separate pile of tacks next to it.

What does this have to do with work automation? Work is constructed into job descriptions similar to the box full of tacks. The job descriptions become a repository of competencies, performance indicators, and reward packages. Soon, leaders, workers, and others see the job and its components as one indivisible thing, like seeing the box full of tacks as one thing. This tendency to think of jobs as fixed repositories obscures powerful opportunities to optimize work automation. It leads to the common but overly simplistic question, "How many workers doing a job will be replaced by automation?" The true pattern of work automation is only revealed in the deconstructed work tasks, not the job.

Just as you must take the tacks out of the box to solve the candle problem, you must take the tasks out of the job and then reinvent the job to solve the work automation problem.

Let's return to the ATM story to see how this works.

The Wrong ATM Question: "How Many Teller Jobs Can Be Replaced?"

Imagine you lead the workforce of a retail bank in the 1970s. Your technology analysts have run the numbers and estimated huge cost savings from replacing the human tellers with ATMs. In fact, because teller machines need not be attached to a full bank branch, your technology planners estimate that eventually you can cut costs even more by reducing the number of full branches, creating mini-branches consisting *solely* of ATMs. Customers who need services beyond the teller machines will go to one of the fewer traditional bank branches. The technologists are also enthusiastic about risk reduction, because teller machines make fewer mistakes, like failing to complete necessary paperwork or coding transactions incorrectly. They wax eloquent about enhancing the customer experience, because ATMs

Work Elements of the Job of Bank Teller

- Greeting and welcoming customers
- Receiving customer's request for cash withdrawal
- Verifying that customer's account balance contains sufficient funds
- Processing the withdrawal to debit the customer's checking account
- Counting and giving cash to the customer
- Counseling customers when account balances are insufficient to process the transaction
- Engaging the customer in conversations
- Detecting customer's receptivity to additional banking services
- Recommending and describing additional banking services
- Referring customers to other bank employees for further services and products
- Collaborating with bank product designers and process leaders to improve products and processes

can process transactions faster so customers spend less time waiting in line. These potential benefits are enticing, but as history shows, simply replacing human tellers with automated machines wasn't the optimal solution.

The first step to a better solution is to take apart, or deconstruct, the job into work elements or tasks. (The sidebar "Work Elements of the Job of Bank Teller" shows one example of how the deconstructed teller job might look.)

Some tasks, such as processing cash withdrawals, are very amenable to the automation of ATMs. Others, such as counseling customers whose accounts have been frozen due to overdrafts, are not amenable to automation. An ATM can hardly deal with customer frustration and anger.

Deconstructing the teller job into its elements also reveals that job elements could be automated in different ways. Human bank tellers assisting a customer completing a simple transaction can detect when that customer might be receptive to other banking services. A recent *Atlantic* article featured an interview with Desiree Dixon, a member-service representative at the Navy Federal Credit Union in Jacksonville, Florida, who described her work: "[W]hen you walk into a Navy Federal, [the staff] really understands what you go through as a military spouse or your family being in the military. Unless you're in that situation, or you have people in relation to that, there isn't that understanding. When your husband or your sister is out to sea and they're deployed, and you're trying to get business taken care of—you may have a power of attorney and it's in their name. Navy Federal really understands that those things occur."[2]

Now you can see more clearly how to group the tasks: some are repetitive (providing requested cash; verifying sufficient funds), while some are variable (collaborating with product designers to improve products and processes). Some require human interactions, empathy, and emotional intelligence (conversing with customers; counseling those who have insufficient funds), while some are done independently (calculating cash balances). Some are physical (giving cash to customers), and some are mental (identifying appropriate additional bank services). You realize that these categories reveal which tasks are very compatible with replacement by an ATM (such as repetitive-independent-physical), and which must be done by humans or automated very differently (variable-interactive-mental). (See table 1-1.)

TABLE 1-1

Work elements categorized by their dimensions

Tasks/work elements of the job of bank teller	DIMENSIONS OF THE WORK ELEMENT		
	Repetitive vs. variable	Independent vs. interactive	Physical vs. mental
Greeting and welcoming customers	Repetitive	Interactive	Mental
Receiving customer's request for cash withdrawal	Repetitive	Interactive	Mental
Verifying that customer account balance contains sufficient funds	Repetitive	Independent	Mental
Processing the withdrawal to debit the customer's checking account	Repetitive	Independent	Mental
Counting and giving the cash to the customer	Repetitive	Independent	Physical
Counseling customers when account balances are insufficient to process the transaction	Variable	Interactive	Mental
Engaging the customer in conversations	Variable	Interactive	Mental
Detecting customer's receptivity to additional banking services	Variable	Interactive	Mental
Recommending and describing additional banking services	Variable	Interactive	Mental
Referring customer to other bank employees for further services and products	Repetitive	Interactive	Mental
Collaborating with bank product designers and process leaders to improve products and processes	Variable	Interactive	Mental

Deconstructing Jobs into Work Elements

As the ATM example illustrates, you must deconstruct jobs into their key elements and not think in terms of replacing entire jobs. Those elements will reveal the optimization patterns, often hidden when the work is trapped in a job description. That does not mean that jobs will disappear, but rather that they will be reinvented, as work that was aggregated into a "job" is constantly reconfigured and continuously deconstructed and reconstructed. Over time, some work elements will be removed from the job as they are transferred to other work arrangements or automation.

The remaining work tasks may no longer make up a full-time job. However, work automation isn't just about optimizing one job at a time. Groups of jobs are related, so work automation requires optimizing the related work tasks across several jobs. In a related group of jobs, each job's content may be reduced by automation, but the remaining human tasks from several related jobs may be combined into a new, reinvented full-time job. Our examples will often focus on a single job for illustration, but you can use the same tools in the more realistic situation where work automation should apply to groups of jobs with related tasks.

How do you find the component tasks within jobs? There are many frameworks. You may be using several of them. You can find the tasks that make up jobs in job descriptions and competency lists. You can also sometimes find them in performance goals and reward components. One online library of work tasks, spanning thousands of different kinds of jobs, is O*Net. Its website says, "[T]he O*NET database, containing hundreds of standardized and occupation-specific descriptors on almost 1,000 occupations covering the entire US economy. The database, which is available to the public at no

FIGURE 1-1

Automation compatibility of tasks within jobs

The impact of automation is best understood by breaking the economy down into tasks

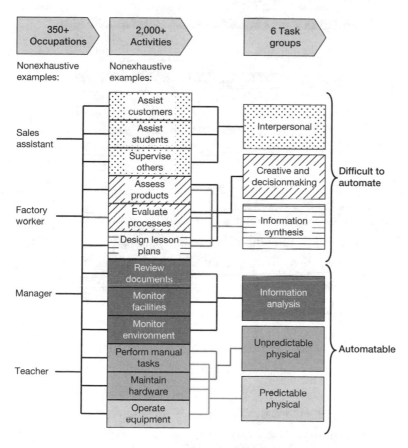

cost, is continually updated from input by a broad range of workers in each occupation."[3] Figure 1-1 is an adaptation of a graphic produced by AlphaBeta Analysis using data from O*Net to illustrate the automation compatibility of tasks within jobs. As you can see, each job contains many different

FIGURE 1-2

Three dimensions that determine automation compatibility

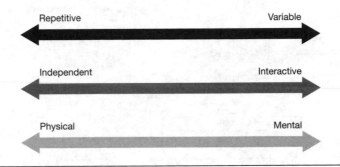

tasks, and each task has different automation compatibility. Asking if a job is compatible with automation is meaningless, compared to asking how automation compatible each deconstructed task is.

What Makes a Task Automation Compatible?

How do you measure the ease of automating a task? We believe there are three fundamental characteristics, as shown in figure 1-2.

Repetitive versus Variable?

Repetitive work is often predictable, routine, and determined by predefined criteria, while more variable work is unpredictable, changing, and requiring adaptive criteria and decision rules.

Most of the work tasks of credit analysts are repetitive. They gather and synthesize similar data for every loan application. They look for the same red flags in each piece of customer data that is pulled from bank records, credit rating agency data, government records, and social media. Generally, repetitive work is more automation compatible with well-established solutions such as RPA, which we describe in chapter 3. RPA can perform such analyses as much as fifteen times faster, with almost no errors. On the other end of the continuum, the work of a human resources consultant is highly variable. Every client situation is different and every problem is unique. This consultant works with analytical tool kits, change management frameworks, and process design techniques that must be customized to diagnose unique problems and solutions. Such work is generally less amenable to automation, but advances in cognitive automation might automate some analytical tasks or learn from previous client engagements.

Independent versus Interactive?

Independent work requires little or no collaboration or communication with others, while work performed interactively involves more collaboration and/or communication with others, and relies more on communication skills and empathy.

Accountants preparing statutory reports for regulators using prescribed templates and decision rules are doing primarily independent work. They can gather data from various sources, synthesize their findings, apply accepted analytical tools, and produce reports with their findings without engaging another person. A good portion of such work is automation compatible using well-established methods. For example, RPA could do the information gathering and synthesis, while

AI could do much of the analysis and produce certain basic reports. Call-center agents, on the other hand, are doing interactive work, matching their work to each caller's unique emotions, needs, and style of communication. Interactive work is generally less automation compatible, but advances in AI and sensors can detect the caller's emotions and analyze the request to give the call-center employee relevant information to better serve the caller with greater empathy and care.

Physical versus Mental?

Physical work is primarily manual in nature, requiring manual dexterity and often strength, while mental work requires one's cognitive abilities.

The work of a manufacturing line assembler is physical work. The assembler might gather different parts, weld them together, inspect the work, and move the finished product to another part of the factory. Such work lends itself well to social or collaborative robotics that is the result of combining AI, sensors, and mobile equipment. A collaborative robot could gather and move parts and weld them together to degrees of precision that greatly exceed the skills of a human being. On the other hand, RPA or cognitive automation can often replace or augment the mental work of an accountant.

Job Deconstruction and Reconfiguration: Oil Drillers

The job of oil driller is at the nexus of massive economic and technological change. Traditionally, the natural resources industry is labor intensive, but cost pressures due to declining commodity prices and margins have demanded greater operational excellence. That is the strategic goal that often motivates technologists and operations research leaders to recommend automating work. And they have made significant advances in automating many aspects of the extraction process. While the strategic benefits of technology are enticing, they rely on deep and radical changes in work and the organization. Technology innovations require work transformations across the entire extraction process. Jobs can be reinvented to reduce physical risk, reduce the probability of accidents with dire consequences, and reinvent work so that it is less demanding and more attractive to increasingly scarce talent.

Let's look at the job of a driller on an oil rig. Much of the work is traditionally repetitive, independent, and physical. In the past, the extensive use of analog equipment emphasized the driller's experience and expertise in ensuring the smooth operation of the rig. As a result of this human centricity, there was significant variation in the performance of each rig. In addition, the driller often did maintenance based on his feel and sense of when equipment might not be operating optimally. Control of the rig was entirely in his hands. The physical nature of the work meant high labor intensity and relatively low skills.

Such work is very automation compatible. Sensors and AI enable a radical reinvention of the work and the driller's job. Now drillers need not be exposed to the elements, physically manipulating equipment on their own. Instead, they sit in climate-controlled cockpits. Their work is to monitor digital gauges that control automated functions on the actual rig. Reinventing the job this way allows for some of the driller's tasks to move to a centralized control center that can monitor multiple rigs at one time by using enhanced sensing equipment and AI that can predict future maintenance events or likely variances in performance. This creates more consistent operating performance. The driller is no longer the only decision maker determining when and how to perform maintenance, because sensors and AI provide specialized maintenance crews with the information to know the optimal schedule and type of maintenance. The job of a driller has been reinvented and is now more mental and interactive. The work is more variable, because automation handles the repetitive parts, saving the human driller for the unique situations.

Table 1-2 shows a sampling of the activities or tasks of the driller following job deconstruction. It classifies the various activities based on our aforementioned categories and assesses whether the work can be performed on-site or at a remote location. Finally, it details the time spent each day on the particular activity.

As a result of the deconstruction of the role, this organization was able to clearly identify how to optimize the application of automation and understand how it would transform various activities. Table 1-3 (page 32) details the output of the deconstruction, automation, and reconstruction of the driller's work. Automation will shift minutes of work to other roles, augment activities, eliminate them, or create new activities.

As you can see from this example, deconstruction is a critical first step to understanding how to apply automation to transform work. But, the exercise is not merely one of deconstructing jobs to identify substitution or augmentation opportunities; the exercise also reveals new work from automation. Analysis begins with understanding the problem to solve. In subsequent chapters, we detail our framework in

TABLE 1-2

Sample deconstructed job tasks of an oil driller

Job name	Performance standard	Activity detail	Activity classification	Possible job locations	Time allocation (in minutes spent)
Driller	Operate within set tripping parameters	The driller maintains open hole conditions	Repetitive, independent, mental	On-site/off-site	10
Driller	Operate within set tripping parameters	The driller does not exceed overpull parameters	Repetitive, independent, physical	On-site/off-site	10
Driller	Operate and use draw works, weight indicator, and auxiliary brake	The driller uses the brake or joystick correctly	Repetitive, independent, physical	On-site	20
Driller	Operate and maintain accumulator, blowout preventer, and choke manifold in accordance with agreed procedures	The driller verifies that the equipment is maintained in accordance with stated procedures	Repetitive, independent, mental	On-site	4
Driller	Operate and maintain accumulator, blowout preventer, and choke manifold in accordance with agreed procedures	The driller can line up the choke manifold	Repetitive, independent, physical	On-site/off-site	4
Driller	Monitor and record trip tank volumes during tripping operations and recognize volume deviation	The driller confirms that the well is being monitored correctly for the operation being conducted	Repetitive, independent, mental	On-site	3
Driller	Manage housekeeping and rig floor organization	The driller leads and directs the crew to maintain housekeeping standards	Variable, interactive, physical	On-site	10

TABLE 1-3

Transformation of the driller's role as a result of automation

	Minutes to perform task (based on 12-hour shift)
Current state of driller's work activities	720
Change due to adoption of AI and robotics	
• Activities shifted *to* other roles	(62)
• Activities *augmented by automation*	82
• Activities eliminated due to automation	(65)
• New activities created due to automation	45
Future state of driller's work activities	720

which the category of "creating new work" reflects two kinds of problem solving:

- Imagining work that cannot be conceived without combining humans and computers.

- The redefinition of the goal as solving the problem because automation allows a close connection between the work and the user's problems. (We will illustrate this in greater detail in chapter 5, when we explore the organizational implications of automation and discuss the intriguing case of Haier.)

A recent article reinforces our idea that the opportunities from automation go beyond the mere substitution of human labor at the task level but instead create opportunities for a more expansive rethinking of work.[4]

We'll now delve deeper into how automation has played out across other aspects of the natural resources value chain and present some case studies. Table 1-4 summarizes

TABLE 1-4

Automation and jobs in the natural resources extraction industry

Phase/job	What's changing?	Case studies
Operation (driller)	Extractive operations can be performed by computer operators who are hundreds or thousands of miles away, requiring a new set of skills to monitor and execute operations (such as hand-eye coordination and advanced cognitive functioning). Ore transportation can be achieved through automated trucks providing greater accuracy, prolonged working time, greater safety, and reduced staffing costs. *Repetitive and physical work eliminated and transformed into mental, variable work.*	• Anglo American has introduced automated drilling in Africa, with good acceptance by workers. Automated drilling brings huge benefits: drill operators can work from a clean, safe, and comfortable command center rather than at a dusty, noisy, and unpredictable mine. • In 2013, BHP Billiton opened an Integrated Remote Operations Centre (IROC) in Perth. The IROC gives the company a real-time view of its entire Western Australia (WA) iron ore supply chain and allows it to remotely control its Pilbara mine, fixed plant, and train and port operations from one central location.
Exploration (geologist, surveyors)	Exploration is modernized using sensors, wireless communication, and computers, which enable greater speed, lower cost, and greater accuracy. *Eliminate repetitive, physical work while augmenting cognitive activity.*	• Freeport-McMoRan uses drones to more closely monitor and evaluate the rock face at mines in real time when blasting away rock to build mine slopes. The drones can see angles that humans cannot see and act objectively. Decisions can be made based on structural data, producing more precise readings and greater productivity.
Processing (quality engineers)	Processing technology increases the efficiency and quality of operations, improving the refining process speed and quality. *Eliminate physical, repetitive work.*	• Metso replaces the work of human inspectors with visual and heat sensors to scan the surface of molten metal to quickly assess steel quality and automatically identify process adjustments that improve product quality.

how automation has changed several jobs in the natural resources extraction industry, with an actual case study of each change. The first example is that of the oil driller; the same patterns emerge in the work of related jobs across the industry, described in the other examples.

Reinventing jobs is a vital factor in connecting work to the strategic organizational goals and operational aspirations of technologists. Pioneer Natural Resources, a US oil and gas producer, achieved the strategic and operational goals of reducing the required days to drill new wells so drastically that it cut costs by 25 percent in wells completed. In 2015, the company added nearly 240 wells to the Permian Basin in Texas without adding one new employee.[5] That required reinventing jobs, as shown in table 1-4. Such reinvention, guided by work deconstruction, is essential to meet the strategic challenges of a highly competitive and cost-pressured environment and the goals of increasing profits and adjusting to price volatility.

What started with simple remote-controlled machines to improve operating control and reduce variance has evolved to encompass integrating work with sensors, automated analytics, and AI-enabled machines that adapt to changing conditions. The work must similarly evolve and be reinvented.

The Long History of Job Deconstruction

In the 1990s, business process reengineering challenged the fundamental underpinnings of specialization in jobs that had characterized organizations for more than a hundred years. In his seminal *Harvard Business Review* article,

"Reengineering Work: Don't Automate, Obliterate," Michael Hammer, the father of reengineering, said,

> The usual methods for boosting performance—process rationalization and automation—haven't yielded the dramatic improvements companies need. In particular, heavy investments in information technology have delivered disappointing results—largely because companies tend to use technology to mechanize old ways of doing business. They leave the existing processes intact and use computers simply to speed them up . . . But speeding up those processes cannot address their fundamental performance deficiencies. Many of our job designs, work flows, control mechanisms, and organizational structures came of age in a different competitive environment and before the advent of the computer. They are geared toward efficiency and control. Yet the watchwords of the new decade are innovation and speed, service and quality. It is time to stop paving the cow paths. Instead of embedding outdated processes in silicon and software, we should obliterate them and start over. We should "reengineer" our businesses: use the power of modern information technology to radically redesign our business processes in order to achieve dramatic improvements in their performance.[6]

What is often overlooked is that the earlier breakthroughs in process reengineering also relied on job deconstruction, work reinvention, and even the integration of work and automation, albeit using far more rudimentary automation tools than we have today. (See table 1-5.) The following example

TABLE 1-5

Reengineering versus deconstruction

	Reengineering	Deconstruction
Focus	Making the organizational silos and jobs work together by reengineering the *process*	Deconstructing jobs into core *work* elements and then reconstructing them to accelerate speed, innovation, and quality
Role of automation	An *enabler* of reengineering by improving information flow and integration across organization silos	A key *driver* of deconstruction and an alternative source of work
Role of strategy	The starting point for reengineering and the basis for rethinking processes	The starting point for deconstruction and the basis for rethinking work
Optimal environment	Ideally suited for environments where the emphasis is on near-term exploitation (versus longer-term exploration)	Relevant to both near-term exploitation and longer-term exploration or innovation

from Hammer's article illustrates this beautifully. Hammer referenced the great success Mutual Benefit Life (MBL) had from reengineering its application process:

> Mutual Benefit Life, the country's eighteenth largest life carrier, has reengineered its processing of insurance applications. Prior to this, MBL handled customers' applications much as its competitors did. The long, multistep process involved credit checking, quoting, rating, underwriting, and so on. An application would have to go through as many as 30 discrete steps, spanning 5 departments and involving 19 people. At the very best, MBL could process an application in 24 hours, but more typical turnarounds ranged from 5 to 25 days—most of the time spent passing information

from one department to the next. (Another insurer
estimated that while an application spent 22 days in
process, it was actually worked on for just 17 minutes.).
MBL's rigid, sequential process led to many complications. . . For instance, when a customer wanted to
cash in an existing policy and purchase a new one,
the old business department first had to authorize the
treasury department to issue a check made payable
to MBL. The check would then accompany the paperwork to the new business department. The president of
MBL, intent on improving customer service, decided
that this nonsense had to stop and demanded a 60%
improvement in productivity. It was clear that such an
ambitious goal would require more than tinkering with
the existing process. Strong measures were in order,
and the management team assigned to the task looked
to technology as a means of achieving them. The team
realized that shared databases and computer networks
could make many different kinds of information available to a single person, while expert systems could help
people with limited experience make sound decisions.
Applying these insights led to a new approach to the
application-handling process, one with wide organizational implications and little resemblance to the old
way of doing business. MBL swept away existing job
definitions and departmental boundaries and created
a new position called a case manager. Case managers
have total responsibility for an application from the
time it is received to the time a policy is issued . . .
Unlike clerks, who performed a fixed task repeatedly
under the watchful gaze of a supervisor, case managers work autonomously. No more handoffs of files and

responsibility, no more shuffling of customer inquiries. Case managers are able to perform all the tasks associated with an insurance application because they are supported by powerful PC-based workstations that run an expert system and connect to a range of automated systems on a mainframe . . . In particularly tough cases, the case manager calls for assistance from a senior underwriter or physician, but these specialists work only as consultants and advisers to the case manager, who never relinquishes control. Empowering individuals to process entire applications has had a tremendous impact on operations. MBL can now complete an application in as little as four hours, and average turnaround takes only two to five days. The company has eliminated 100 field office positions, and case managers can handle more than twice the volume of new applications the company previously could process.[7]

Notice how the strategic goals that motivated reengineering (cost, reliability, and efficiency) require reinventing the job of case managers and the other related jobs. Notice how process reengineering required reinventing the job by first deconstructing it and then moving some parts to automation (PCs and early databases), keeping other parts as they were, and adding new work that requires taking full accountability of the case process.

The point is that virtually all organizations have used process reengineering for a long time. That very likely required optimally deconstructing and reinventing jobs. Today, such strategic reinvention could use more advanced automation tools. We might use RPA and AI for most of the data gathering, analysis, and processing, leaving the case manager to review the automated recommendations. Instead of building expensive databases and networks to integrate all data into a single source, the combination of RPA and AI could automatically gather data from multiple independent data sources and apply pattern recognition to analyze structured and unstructured data through natural language processing.

However, whether in process reengineering or automation optimization, the fundamental role of work deconstruction and reinvention is very similar. If your organization has done process reengineering, it has very likely done job deconstruction and reinvention. Now, you can tap those capabilities in service of optimizing work automation, just as they were used to optimize process reengineering.

Deconstructing jobs into work tasks reveals the essential work patterns to optimize automation. The 2017 Willis Towers Watson study of the future of work identified deconstruction as one of the top two future opportunities to enhance organizational readiness for automation. However, deconstruction to identify automation compatibility is just the start. A second vital question asks what payoff work automation can produce. That question takes those same deconstructed job tasks and identifies the value of improved performance. That's the topic of the next chapter.

TWO

Assess the Relationship between Job Performance and Strategic Value

What Is the Automation Payoff?

Pivotal strategic goals, like agility, customer responsiveness, cost control, and innovation drive automation decisions. Often, leaders assume that automation itself produces those results, but, in fact, they depend on execution. Execution happens through the work and how it's organized. That's obvious, but is improving all work performance equally valuable? No. Improving certain pivotal areas of work performance has an outsized impact on strategic goals, while improving other areas of work performance has little effect.

As John Boudreau and Peter Ramstad explained in their book *Beyond HR*, optimizing the connection between strategic goals and work requires understanding how improved work performance actually adds value and requires investing in work performance based on that value. Unfortunately, most strategies stop short of this deep understanding about work, relying instead on generic investments in people, such as "get me the best person for every job" or "create a culture where everyone strives for their highest performance." Such generic approaches produce generic work improvements that are not targeted to strategic goals.

The same thing applies to work automation. Automation can only achieve its strategic goals when organizations invest in where it produces the most pivotal payoffs. Those payoffs uniquely reflect the organization's strategic goals, resources, processes, and culture. Generic work-automation investments are wasteful and potentially harmful.

So, reaping the benefits and mitigating the risks of your strategy requires addressing a new set of challenges:

- Should you automate all the work elements in an entire job? That would follow the typical idea of "robots replacing workers."

- Should you automate the work elements that take up the most time? That would cut labor costs but might reduce productivity and increase risk if those time-consuming tasks need a human touch.

- Should you automate the work elements where humans are less capable? That would follow the idea that people should do the work that is most "human," but mundane work may be more productive, less risky, or less costly if done by humans.

None of these simple challenges adequately captures the required trade-offs, because not all work tasks pay off in the same way. Some tasks yield a high return only when performed very well. Other tasks damage company value and brand when they fall below a certain standard. Still others create no difference in value even though they are done in many ways.

How can we more clearly understand these differences and the potential role of automation in optimizing performance? The key is to clearly define the relationship between work performance and the value it creates, the "return on improved performance" (ROIP).

Consider the job of science director in the pharmaceutical industry. If you ask, "What is the value of better job performance?" you can't really answer the question because the job contains so many elements. One work element is research, where performance can range from moderate (being aware of leading research) to great (being a big thinker who publishes breakthrough research). Another work element is team leading, where performance can range from moderate (providing input to the team) to great (creating collaboration that transforms breakthrough ideas into unique drug formulations). The value of different tasks depends on the strategic goals. For organizations that already have lots of breakthrough ideas, the strategic goal of innovation is most improved by a research scientist who is great at team leading and moderate at research. Conversely, when an organization already has team processes that are running well, the strategic goal of innovation is most improved by a research scientist who is great at research and moderate on team leading.

ROIP applied to deconstructed work tasks reveals the payoff patterns that connect work performance to strategic goals. Let's return to the ATM example to see how this works.

ATMs, Bank Tellers' Work, and ROIP

Deconstructing the teller job into its tasks has revealed more options than you or your technologists imagined when you thought simply to "replace tellers with ATMs." But now you have a new dilemma. Choosing among the options requires identifying which options are most pivotal, how they pay off, and what should be your priority.

Work performance pays off in different ways, and that varies among the tasks in a job. Some tasks create value by avoiding mistakes, but not performing beyond that standard. Verifying that an account has sufficient funds to cover a withdrawal is vital, but once you know funds are sufficient, there is little additional value in being more precise about the available amount or type of funds. Other tasks create additional value from each incremental performance improvement. When recommending additional banking services, each increment in the accuracy and enthusiasm of the teller adds additional value. For some tasks, different ways of performing don't create a different payoff. Conversing with customers in a pleasant and friendly way can be done in many ways, but creates the same value.

If you apply the ROIP idea to your ATM dilemma, you might identify different payoff functions for the different work elements (see table 2-1).

Automating some tasks will reduce risk, automating others will incrementally increase quality, and automating others will reduce variance that adds no value. Each type of ROIP implies a different payoff. ROIP connects your original strategic goals to actual task performance, so now your payoff analysis can be more precise.

TABLE 2-1

ROIP for bank tellers' tasks

Work elements of the job of bank teller	Return on improved performance (ROIP)
Greeting and welcoming customers	Many different ways, same value
Receiving customer's request for cash withdrawal	Avoid mistakes
Verifying that customer's account balance contains sufficient funds	Avoid mistakes
Processing the withdrawal to debit the customer's checking account	Avoid mistakes
Counting and giving the cash to the customer	Avoid mistakes
Counseling customers when account balances are insufficient to process the transaction	Avoid mistakes; very high performance can salvage a potentially lost customer
Engaging the customer in conversations	Many different ways, same value
Detecting customer's receptivity to additional banking services	Better performance produces incrementally more value
Recommending and describing additional banking services	Better performance produces incrementally more value
Referring customer to other bank employees for further services and products	Many different ways, same value
Collaborating with bank product designers and process leaders to improve products and processes	Little danger of damage from poor performance, moderate value of medium performance, very high value for exemplary inventiveness

So, ROIP is also key to optimizing strategic work automation. If we applied automation to the research scientist job, being moderately good at research means being aware of newly published findings, so an automated research alert provides the optimal payoff. Here, automation reduces the chance of missing an important publication, though it won't increase unique breakthroughs. On the other hand, being great at research means generating unique new break-throughs, so that outcome will require different work auto-mation using highly advanced AI that can observe and interact with the research scientists, learning the patterns that lead to unique breakthroughs. ROIP defines the payoff from the work tasks, which determines the payoff from work automation and its contribution to strategic goals.

Four Fundamental ROIP Curves

ROIP can take many forms, but we can illustrate the power of the idea with four prototypical ROIP relationships. We use tax preparation as our example, noting that these patterns exist in virtually all work.

In figure 2-1 (page 47), the vertical axis represents the value of work performance to the organization, and the hori-zontal axis represents the level of performance. The height of the curve shows the ROIP of a particular performance level. The ROIP curve is divided into four sections, depicting four prototype curves.

Negative Value: Reduce Mistakes

The left section of the figure shows the return on improv-ing performance ranging from very low levels that gener-ate negative value to the minimally acceptable level. The payoff to performance improvements in this range is from reducing the negative value. For tax form preparation, this range would include performance at a very low level, with many mistakes or missing deadlines. The top of this range

FIGURE 2-1

Step 2: Assess return on improved performance (ROIP)

ROIP for full range of potential work performance

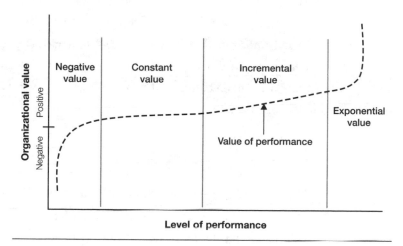

Level of performance

is not great performance, but rather a minimally acceptable performance level that generates a small positive strategic value. For tax preparation, this would be completing forms correctly and on time.

Constant Value: Reduce Variance

The second section of the figure shows the ROIP range where better or worse performance actually makes little impact on strategic value. Performance differences in this range all produce the same value. For tax form preparation, this range would include completing the form at any time before or at the due date. Better performance (getting the form completed far ahead of the due date) adds no more value than completing the form just in time for the due date. This range could also reflect differences in accuracy that have no material effect on the tax outcome, such as the difference between calculating values to the penny when the form requires accuracy only to the nearest dollar. The constant-value curve often reflects work tasks with many ways to reach the goal, such as when different workers might assemble components in different sequences, but the final assembly is essentially identical. Or, different call-center representatives might use a customer's name between one and three times, but that makes no difference to the customer's satisfaction with the service.

Incremental Value: Steady Improvement

The third section of the figure shows the ROIP range where each performance improvement produces a steady incremental increase in strategic value. In tax form preparation, this ROIP range might reflect performance on the clarity and grammar quality of the summary letter that accompanies a client's tax form. A minimally clear letter satisfies the minimum requirement, but if the letter is more clearly written and/or points out more important highlights, that is incrementally valuable to the client and the organization. Another example would be when a call-center representative upsells customers in incremental ways by suggesting additional features or faster shipping.

Exponential Value: Breakthrough Performance

Finally, the right section of the figure shows an ROIP range where improved performance increases value exponentially. This range often represents very rare or creative performance that surprises and delights a customer or disruptively improves a process. In tax form preparation, this ROIP range might reflect when a tax preparer discovers an obscure tax deduction or a very intricate method of restating income to significantly reduce owed taxes. Another example would be when an in-store retail associate or call-center representative uncovers obscure customer information that reveals an unusual need for products or services that carry a much higher value to the organization. In this performance range, breakthrough innovations have huge payoffs, such as when biochemists make rare and pivotal disease treatment discoveries or when social media productions achieve viral popularity.

ROIP and Strategic Performance on McDonald's versus Starbucks's Front Line

In *Beyond HR,* John Boudreau and Peter Ramstad suggested that the ROIP can explain how the different strategies of McDonald's versus Starbucks show the different performance value of in-store service associates.[1] Their job descriptions are similar. Both roles involve interacting with customers, taking payments, working with the team, having reliable attendance, and correctly preparing and delivering the orders. But ROIP analysis reveals hidden and important strategic differences.

McDonald's and Starbucks choose to compete differently. McDonald's is known for consistency and speed. Its stores automate many of the key tasks of food preparation, customer interaction, and team roles. McDonald's has assigned numbers to its products, so associates need only press the number on the register to place the customer order and calculate the price. Indeed, McDonald's customers often order by saying, "I'll take a number three with a Coke and supersize it."[2] This is a good strategy for McDonald's, because it can acquire and deploy a wide variety of talent in its stores. The work design minimizes mistakes. However, the chance for significant performance breakthroughs is also low. The ROIP range for McDonald's associates is negative value and constant value.

Contrast this with Starbucks, where baristas are a multi-talented group. The allure of Starbucks as a "third place" to hang out and work is in part predicated on interesting, extended interactions with Starbucks baristas. Some are opera singers who sing out the orders. Their styles are clearly on display and range from Gothic to country to hipster. Star-bucks counts on that diversity as part of its image.[3] That means it gives its baristas wide latitude to sing, joke, and kib-itz with customers. The ROIP range for Starbucks is wider and includes incremental value and even exponential value, but also negative value. There is more upside performance value but also more room for error at Starbucks. That room for error is the price that Starbucks pays to gain the opportu-nity for breakthrough value. (See figure 2-2 [page 51].)

McDonald's designs the work to maintain very tight distri-bution (shown by the bottom arrow in the figure). It guards against poor performance by restricting the performance range because its strategy does not require creative service on the high end of the performance scale. McDonald's wants high performance on the job, but it defines "high" within a narrow range. Starbucks accepts and even encourages a wide array of performance levels, because the way it competes cre-ates a high payoff from an extraordinary innovation (the top arrow in the figure). Starbucks also wants to guard against poor performance, but if it wants the high end, it may have to accept some risk of the low end. For example, some customers

FIGURE 2-2

ROIP analysis for frontline work: McDonald's versus Starbucks

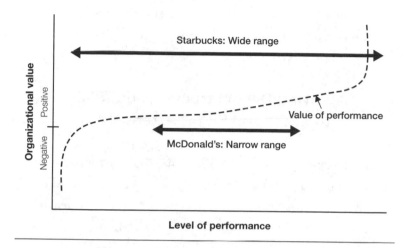

may be annoyed if the opera singer gets too boisterous, but Starbucks can't simply say "no singing," as McDonald's can. Starbucks must allow the singer to have a chance to delight the customer.

How does this reveal different work-optimization solutions at McDonald's versus Starbucks? McDonald's extensively uses automated ordering systems and kitchen process automation that minimize variations, achieving the same outcome with limited choices. Starbucks also has automated ordering systems, but it must encompass a vast array of customized preferences, relying on the human associate to interpret and execute them. When Starbucks introduced systems that allowed customers to order online before they arrived at the store, some in-store customers resented that their barista was so busy filling those online orders that he or she didn't have time to chat. Starbucks's automation must balance efficiency and scale very differently with customer interaction. Indeed, breakthrough automation at Starbucks might use algorithms to recall and communicate customer details like favorite music, children's names, and so on, so that associates can have immediate familiarity with repeat customers.

Applying Deconstruction and ROIP to Pilots and Flight Attendants

We use the example of airline pilots and flight attendants here and in chapter 4 to show how the steps in the framework combine. We'll apply work deconstruction and ROIP to these jobs.

Pilots are a critical talent pool for an airline, and operating the aircraft is a vital work task. Let's apply the ROIP analysis to that task. In figure 2-3 (page 53), the task of operating the aircraft has an ROIP curve range from negative value to constant value. Operating the aircraft at a very high standard achieves full value. Higher performance will not yield additional strategic value, but having even one pilot perform at below minimum standards can have a significantly negative impact. This is the reason airlines invest in elongated career paths for pilots. For instance, it takes twenty years to move from the right seat of an Embraer 175 doing a short-haul flight to the left seat of a Boeing 747 going across the Pacific Ocean. Airlines significantly invest in cockpit technology as well as in training and development (e.g., minimum simulator hours required) in order to ensure that performance does not fall to the left of standard.

Now, compare the ROIP curve for pilots to that for flight attendants, also shown in the figure. As airlines pursue value by differentiating the customer experience—particularly for

FIGURE 2-3

ROIP for pilots versus flight attendants

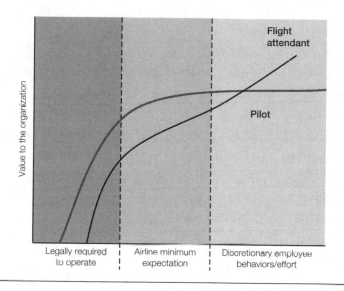

premium passengers—flight attendants are the face of the airline. Flight attendants achieving higher performance on the work element of serving passengers add great strategic value. Flight attendance performance on the work element of serving passengers includes the full ROIP range. In the middle range of performance, this is an incremental-value ROIP curve. At very high performance levels, such as delivering an unusual service that truly delights a passenger, the airline achieves the exponential-value ROIP. Flight attendant performance also includes the negative-value ROIP range at the left of the figure, because attendants must avoid mistakes. The left side of the flight attendant ROIP curve is very much like that for pilots. However, unlike pilots' performance, the improved performance of flight attendants, beyond the minimum standard, adds significant value.

Deconstruction, ROIP, and Work Automation

Deconstructed work elements, combined with the four ROIP curves, reveal deeper insights about work and automation optimization beyond job descriptions that assume incremental ROIP for all work tasks. However, that's just the beginning.

In the next chapter, we provide a framework that describes several categories of automation. When we apply these automation categories to the four different ROIP curves, we reveal new work automation options, and the relative value of those options is clearer. The framework reveals better and more nuanced answers to questions such as: "How can we apply the rapid advances in artificial intelligence to further enhance the impact of these roles?" "Should work automation merely reduce labor costs, or should it increase the performance of the human workers, or both?"

You can imagine lots of work automation options, such as giving flight attendants AI-enabled glasses that display the passengers' names, preferences, and history with the airline. Imagine flight attendants wearing a version of Google Glass, through which they can access customer data and personalized preferences. No nut dishes served to Charles in 3C given his allergy, but black coffee and a predisposition for onboard duty-free purchases. Early seating meal for Sarah in 2A so she can get to sleep quickly. That might enable attendants to deliver breakthrough service along the exponential-value ROIP curve. On the left side of the performance curve, the airline might install automated sensors that detect unfastened seat belts or bags blocking the aisle, which would more reliably ensure that it meets minimum safety and legal requirements. This would reduce flight attendants' mistakes in the negative-value ROIP range and free them to focus on highly personalized service in the exponential-value ROIP range.

For pilots, social robotics might change the negative-value range of the ROIP curve. The aircraft could become a remotely piloted collaborative robot, with in-cockpit pilots replaced by a highly skilled human pilot who oversees multiple flights from an air-traffic control center, intervening when an unforeseen event requires work beyond the capabilities of the robotic pilot. This would allow leveraging the experience and insight of the shrinking pool of skilled pilots much more efficiently. The net effect is a reduction both in labor cost (as fewer pilots are required) and in the risk of an accident.

Once you've done the first two steps—deconstructed jobs into tasks and analyzed the ROIP for those tasks—you can better choose among the AI options. That requires unpacking the different types of automation (chapter 3) and then combining ROIP and automation to reinvent and optimize the work (chapter 4).

- - - - - - - -

THREE

Identify Options

What Automation Is Possible?

When making automation decisions, most leaders begin with this third step. But you can't really choose among the options and make decisions until you've first considered the work elements and the ROIP. Once you've done that, it's easier to see which automation options are best. Let's go back to the ATM example to see how that works.

Optimizing Bank Work Automation

More precise work tasks and payoff functions tell you which work to automate and why, but you still need to decide *how* to automate. That requires identifying the different kinds of automation and their applicability.

Work and automation change almost daily, so any framework for describing automation is by definition incomplete and must change over time. We will analyze automation using the following three widely accepted automation categories:

- Robotic process automation: used for high-volume, low-complexity, and routine tasks. Particularly effective where data needs to be transferred from one software system to another, but requires no learning from interactions.

- Cognitive automation, AI, machine learning: used for nonroutine, complex, creative, and often exploratory tasks. Particularly effective in recognizing patterns and understanding meaning in big data, and where learning from interactions is required.

- Collaborative or social robotics: used for collaborative tasks, for both routine and nonroutine tasks. Robots are mobile and move around our everyday world; they are programmable and adapt to new tasks.

Each of the automation types fits a different kind of work task and provides a different kind of payoff. Now, with the job deconstructed into tasks, the ROIP identified, and your automation options described, you can put it all together to optimize automation for each task, to achieve the proper payoff, and then reinvent the job and its organization and leadership context.

Reinventing the Job to Optimize Work Automation

In a bank teller job, some of the work tasks are very repetitive and require little thinking (counting cash) and are ideal for robotic process automation. Others are highly variable and require a great deal of thinking (collaborating with product designers), where AI could enhance the process. Some involve substituting automation for humans (e.g., verifying account balances), while others augment humans (e.g., recommending that a human do the banking services, but AI can identify the best services and make appropriate recommendations to the human). A routine process (counting and giving cash) can accomplish some tasks, while other tasks are best accomplished by automating the cognition to detect patterns, identify the best option, or make recommendations (predicting customer receptivity to additional services). In yet other areas, automation can create new types of work (e.g., supporting customers remotely as they engage with the bank without ever walking into a branch).

Now you see more clearly how some automation applies better to work elements with certain characteristics and ROIP. You can now more precisely define the cost-quality-risk implications of the combinations of tasks, ROIP, and automation. Table 3-1 (pages 60 and 61) shows how you might define the optimal work automation combinations.

You can clearly see a reinvented job that used to be that of a bank teller. As your technologists predicted, a group of tasks is optimally automated by substituting ATMs (process automation) for humans. However, such tasks are only a subset of the reinvented bank teller "job." For many work elements, human workers remain the optimum solution. For still other tasks, the reinvented work will combine humans and automation.

TABLE 3-1

Optimal work-automation combinations

Tasks or work elements for bank tellers	Return on improved performance	CHARACTERISTICS OF THE WORK ELEMENT			Impact of automation: Substitute, augment, create	Automation type
		Repetitive vs. variable	Independent vs. interactive	Physical vs. mental		
Greeting and welcoming customers	Many different ways, same value	Repetitive	Interactive	Mental	Augment	Cognitive automation
Receiving customer's request for cash withdrawal	Avoid mistakes	Repetitive	Interactive	Mental	Substitute	Process automation
Verifying that customer's account balance contains sufficient funds	Avoid mistakes	Repetitive	Independent	Mental	Substitute	Process automation
Processing the withdrawal to debit the customer's checking account	Avoid mistakes	Repetitive	Independent	Mental	Substitute	Process automation
Counting and giving the cash to the customer	Avoid mistakes	Repetitive	Independent	Physical	Substitute	Process automation
Counseling customers when account balances are insufficient to process the transaction	Avoid mistakes; very high performance can salvage a potentially lost customer	Variable	Interactive	Mental	Augment	Cognitive automation

Engaging the customer in conversations	Many different ways, same value	Variable	Interactive	Mental	Augment	Cognitive automation
Detecting customer's receptivity to additional banking services	Better performance produces incrementally more value	Variable	Interactive	Mental	Augment	Cognitive automation
Recommending and describing additional banking services	Better performance produces incrementally more value	Variable	Interactive	Mental	Augment	Cognitive automation
Referring customer to other bank employees for further services and products	Many different ways, same value	Repetitive	Interactive	Mental	Augment	Cognitive automation
Collaborating with bank product designers and process leaders to improve products and processes	Little damage from poor performance, moderate value from medium performance, very high value for exemplary inventiveness	Variable	Interactive	Mental	Augment	Cognitive automation
Customer assistance and technology support	Better performance produces incrementally more value	Variable	Interactive	Mental	Create	Cognitive automation

FIGURE 3-1

Automation in the financial service sector

Over the next 20 years, work in the financial services industry is considered at high risk of automation, more than any other skilled industry. About 54% of all work may be eliminated.

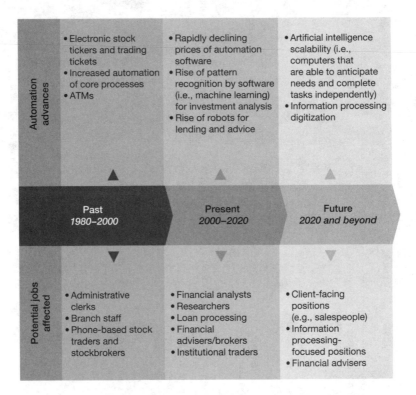

Source: Willis Towers Watson analysis; and Nathaniel Popper, "The Robots Are Coming for Wall Street," *New York Times Magazine*, February 25, 2016.

Let's now look in depth at how automation has and will further extend into the financial services sector (see figure 3-1).

We explored how ATMs affected work in bank branches. When they first arrived in the 1970s, ATMs were expected to spell doom for bank tellers by taking over some of their routine and repetitive tasks. Indeed, in the United States, the average number of tellers fell from twenty per branch in 1988 to thirteen in 2004 to currently fewer than five in some branches. While this decline reduced the cost of running a bank branch, it allowed banks to open more branches in response to customer demand. The number of urban bank branches rose by 43 percent over the same period, so the total number of tellers actually increased. As we explained in chapter 1, rather than destroying jobs, ATMs changed bank employees' work mix away from repetitive tasks and toward things like sales and customer service that machines could not do. As we move forward in time, we see automation shifting from a focus on transforming core transactional processes toward the application of intelligence to higher-valued-added activities like trading and analysis. In the future, even more nonrepetitive, highly cognitive work will either substitute for or augment human capability as AI moves from a focus on the "known unknown" (e.g., I know I don't know what the optimal asset mix should be for someone of my age and risk tolerance) to the "unknown unknown" (e.g., I don't know that I will need to reallocate my portfolio in the event of an unrelated market event, but AI will anticipate that and execute the transactions needed to achieve the right mix of assets).

As we expand our horizons beyond financial services, what various types of automation are available and how will they develop into the future? (See the sidebar "Why Is AI a Big Deal Now? Convergence.")

The Three Forms of Automation

As we mentioned earlier, work automation technologies fall into three categories: robotic process automation, cognitive automation, and social or collaborative robotics (see table 3-2 [page 65]).[1] Their effects on work can be distinguished by:

- Different work tasks that they can automate

- Different ways of learning from and interacting with people

- Different application types and scope

- Different maturity levels

- Different implementation and maintenance costs

- Different implementation time

- Different levels and types of returns

We define each automation category with examples of its effects on work. These definitions and examples identify how the three categories of automation converge to affect work in your organization.

Robotic Process Automation (RPA)

RPA is the simplest and most mature category. RPA automates high-volume, low-complexity, and routine tasks. For example, it has long been used to automate "swivel-chair" tasks that used to require a person to swivel from one data source to another to transfer or connect data from disparate systems. A common application involves transferring data between software systems or using simple rules to find information in emails or spreadsheets and entering it into business systems like enterprise resource planning (ERP) or customer relationship management (CRM). These tasks are often too simple for a complicated IT solution. Instead,

TABLE 3-2

Step 3: The three types of automation

	Robotic process automation	Cognitive automation	Social or collaborative robotics
Work task automation	High volume, low complexity, routine	Complex, exploratory, non-routine, decision supporting	Mixed routine and nonroutine; collaborative
Learning and interacting	Instruction-based; likely to be further enhanced with machine learning	Machine learning, deep nets, hybrid AI; needs data and human trainers to learn	Learning from human interaction and data
Application type and scope	Wide; can automate tasks of business processes	Focused; targeted to specific data sets and tasked to deliver specific outputs (no artificial general intelligence yet)	Wide; can leverage human productivity across a spectrum of activities and expertise
Maturity levels	Maturing, off the shelf	Emerging, with some ready to use (e.g., image or speech recognition)	Maturing, off the shelf
Implementation and maintenance cost	Low	High	Medium/high
Implementation time	Weeks	Months	Months
Level and type of returns	High; can fit the current operational and business models; can reduce need for some offshoring	High; potential to transform operational and business models	High; can significantly enhance productivity and efficiency

simple process robotics can automate them quickly and cheaply, without requiring the management and training of labor. Xchanging, a UK-based insurance claim services company used twenty-seven Blue Prism robots to automate

Why Is AI a Big Deal Now? Convergence

It has often been said that the future is here but unevenly distributed. The same is true for automation. The World Economic Forum graph shown here illustrates why the "fourth industrial revolution" is fundamentally different from the second and third: convergence. Yes, the internal combustion engine and the light bulb were invented within twenty years of each other during the second industrial revolution, but multiple game-changing technologies did not emerge and build on each other, as is happening now. The graph shows how various technologies have matured over time and achieved their cumulative capability (defined as maximum utilization and value realization). Legacy technologies from the third industrial revolution, like mainframe computers and PCs, achieve their maximum utility and then decline as alternative technologies usurp their functionality. But in the fourth revolution, multiple technologies (mobile, big data, and the Internet of Things [IoT]) converge and reinforce each other. At no time have so many technologies approached their peak at the same time (the gray oval in the figure). Convergence is occurring on two dimensions: convergence of the various spheres of our own lives (work, social, biological), and convergence across industries. In the former, common technologies (social media, mobile, sensors) cut across all the spheres of our existence and bring them together. AI and sensors powering the apps in our phones and watches are increasingly the epicenter of every facet of our existence. In the latter, every business is digital and global. The size or the nature of output does not matter, whether of a global bank, a large automotive manufacturer, or a local retailer; the ability to tap the global supply chain, conduct commerce in the cloud, and incorporate personal digital devices will ulti-mately determine strategic sustainability.

As we look at the evolutions of various technologies, how does AI fit into the picture? Is it independent or an accelerant, in much the same way big data and IoT are mutually reinforc-ing? AI is an accelerant in that it allows existing technologies to enhance their value. Mobile and cloud (Web 2.0) coupled with cheap, miniaturized sensors enable the IoT by seam-lessly connecting machines worldwide, gathering, storing, and monitoring huge volumes of data at minimal cost. And AI enables that data to turn into insight and intelligence so that it is rapidly becoming the primary source of competitive

Technology convergence

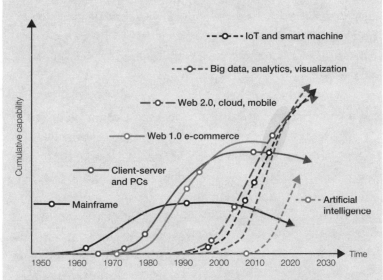

Source: World Economic Forum, "Onward and Upward? The Transformational Power of Technology," 2016, http://reports. weforum.org/digital-transformation/onward-and-upward-the-transformative-power-of-technology/.

advantage by enabling the transformation of virtually every process and value exchange with customers. So the key to the power of automation to transform work is not just the individual capabilities of the technologies, but rather the transformative capabilities that arise from their combination.

The Three Rs of RPA

Any task that one or more people perform over and over again, with little variation, will likely be performed far better by RPA. RPA can be counted on to perform repetitive tasks much more reliably than people can. If you need the same result from the same inputs, with near-perfect repeatability, RPA is your answer. The fact that RPA can perform these tasks around the clock without tiring further increases its value.

The second category of tasks that are ideal for RPA are redundant steps in business processes. Despite decades of business process optimization, most processes are still filled with redundant, nonvalue-added steps. When processes require a person to review, approve, check, audit, supervise, or confirm something, there is a redundancy crying out for robotics. Many redundancies originated in paper-based business processes and have survived attempts at automation. Humans make mistakes, and the redundancies were designed into the processes to catch and correct such errors. Such processes survive because of habit, expectation, regulation, or fear. Each of the factors is fairly sticky, which explains why process redundancies survive. This leads to interesting questions: Who should be the maker and who should be the checker in such processes? If RPA can apply rules perfectly, what is the value of having a human check its work? Conversely, would it make sense for a human to be a maker whose work is checked by RPA? Might the best checker of an RPA's work be another RPA, and would this meet the requirements of rules or regulations? These questions will likely cause your organization much discomfort. Innovation is characterized by discomfort and uncertainty. If your organization isn't struggling with such doubts, you're not innovating.

The key message of redundancy and RPA is to automate both horizontally *and* vertically. While there is value in replacing doers, there may be far more to be gained in replacing watchers, too.

Finally, the third category where RPA may shine is in managing risk, particularly regulatory risk. Many businesspeople may be convinced that regulatory oversight demands that humans oversee their processes. This may simply be a matter of interpretation. As mentioned earlier, the point of many redundant process steps is to ensure that rules have been followed or conditions have been met. Who better to apply these rules or ensure such criteria are met than an unemotional, unbiased, inexhaustible, unforgiving block of software code? One could argue that such reviews or audits are *best* left to RPA, rather than frail, fallible, forgiving, or forgetful humans. For many businesses, process redundancies are regulatory artifacts, and they're notoriously expensive to maintain. Typically, such oversight is required by regulations that were crafted decades ago. Compliance departments and risk managers are often very skeptical of change. After all, they are paid to be paranoid.

> But the consistency with which RPA performs tasks can greatly simplify compliance obligations, generating substantial reductions in operational risk.
>
> *Source*: Adapted from Christopher Surdak, "Robotic Process Automation 101 (Part 2: Where and When)," Institute for Robotic Process Automation and Artificial intelligence, https://irpaai .com/robotic-process-automation-101-part-2/.

fourteen core processes, performing 120,000 RPA transactions per month and reducing the cost per process by 30 percent.[2] (See the sidebar "The Three Rs of RPA.")

A typical RPA algorithm would look something like the following:

Log on system

Open xls file

Copy first three values from column "date of birth"

Open Word document

Paste values on page 3 under heading "date of birth"

Close Word document

Open email

Attach Word document on email

Cognitive Automation

The current headlines about work automation reflect *cognitive automation*, which replaces humans doing nonroutine complex tasks, literally automating human cognition. Cognitive automation uses tools like pattern recognition and language understanding. The retailer Amazon pursued strategic goals that include improving the quality and reducing the cost of customer service in physical stores. That resulted in the Amazon Go retail store in Seattle, which has no cashiers or checkout lanes. Customers shop and go, as sensors and algorithms automatically charge their Amazon account. These strategic goals and operational designs rest on reinventing jobs. Automation does the tasks of scanning purchases and processing payment. This doesn't mean the end of store associates, but their work changes. Humans still do tasks like advising in-store customers about product features. Cognitive automation in the form of machine learning, using scalable cloud computing resources, has produced systems that can recognize patterns and understand meaning in big data in a human-like way. This recognition intelligence is a combination of artificial intelligence, specifically machine learning, and sensors. It is at the heart of automating tasks like voice and image recognition, voice conversion to text, and natural language understanding.

These applications reflect automation that is taught rules and procedures by humans, but in newer, deep learning, the machines teach themselves. This automation is applied to increasingly more diverse, abstract, and advanced tasks. The Google DeepMind team created a computer called AlphaGo that famously defeated master players at the complicated game of Go. To train AlphaGo, DeepMind fed the system thousands of games that amateur and professional human Go players had played. AlphaGo used the games to develop winning strategies and identify good and bad moves. More recently, the same DeepMind team created AlphaGo Zero, a computer that played only by itself (millions of times), at first making moves at random until it recognized strategies. AlphaGo Zero got its name because it had zero help from humans beyond starting it up. AlphaGo Zero defeated not only human players, but ultimately even its predecessor, AlphaGo.[3]

Cognitive automation is typically used in three ways. First, to transform business processes, such as car insurers that use an app with image recognition and cognitive analysis capability to process photos of a damaged car, assess the damage, estimate the size of the claim, and send its recommendation to a human assessor for final approval, creating a simpler, faster, and cheaper claims process. This reinvents the former job of human field inspectors to a job of remote, high-level approvers and assessors. Such technology allows traditional jobs to be deconstructed and to augment or replace routine human activities with automation, resulting in the work being reinvented for greater efficiency, effectiveness, and impact.

Second, cognitive automation can develop new products and services. The same automation that reinvented the claims process enables a new service offering to car insurance clients, with features such as a chatbot that provides on-demand advice about repairs and payments to policy owners, right on their phone. Now, the jobs of customer service associates can and must be reinvented.

Third, cognitive automation can gain new insights with big data. In the auto insurance example, cognitive automation can analyze thousands of claims to identify locations most prone to accidents and compute customer premiums that adjust for driving in high-risk versus low-risk locations. Now, jobs such as data scientists and analysts can and must be reinvented.

You can see how convergence creates exponential automation opportunities. RPA is often a precursor to using AI, where RPA produces the necessary, clean, high-volume data needed to drive effective cognitive automation. Consider the previous RPA illustration with cognitive automation inserted:

Log onto system

Open email

Read email (cognitive AI with Natural Language Processing capability)

If email content requires a list of dates of birth, find the relevant xls file

Open xls file

Copy first three values from column "date of birth"

Open Word document

Paste values on page 3 under heading "date of birth"

Close Word document

Open email

Attach Word document to email

This same convergence applies to work, creating similarly exponential opportunities and requirements to reinvent the work and its organization. Ultimately, optimizing work automation is an opportunity to consider an entire ecosystem of jobs and their relationships, deconstruct them, assess the ROIP of the work, apply RPA and cognitive automation, and then reinvent them. (See the sidebar "Uptake Keeps Trains Running Using Cognitive Automation.")

Social or Collaborative Robotics

You may think of robots as machines bolted to the floor of an assembly line, performing one repetitive task. That's

Uptake Keeps Trains Running Using Cognitive Automation

Uptake, an industrial analytics company, creates "purpose-built products [that] ingest and analyze sensor and enterprise data, transforming it into actionable insights and immediate outcomes." CEO Brad Keywell relies on cognitive automation and reinvented jobs, such as maintaining railroad locomotives. Railroad locomotives are powered by massive, highly complex electrical engines that cost millions of dollars. A breakdown costs the railroad thousands of dollars for every hour out of service, plus customer irritation and frustration. In the past, a broken locomotive was towed into a repair facility. Only at that facility could technicians run diagnostic tests, which took hours.

A strategic goal for rail operators was to use automation to reduce breakdowns and more effectively and efficiently maintain locomotives. That requires automation, but it must be seamlessly integrated with a reinvented repair mechanic job. Uptake's AI and algorithms constantly run diagnostics on the operating locomotives, long before they break down. Uptake's systems predict when, why, and how the locomotive might break down. They use predic-

still true, but increasingly giving way to *social robotics*. The word "social" refers to robots that move around and interact with people, using sensors, AI, and mechanical machinery. A subset of social robotics is "collaborative" robotics (cobots). Cobots are machines that actually sense the human worker and actively adjust to physically work with the human.

tive analytics, powered by cognitive automation analyzing massive data generated by 250 sensors on each locomotive. Uptake systems analyze the data using the operating history of similar machines, training from human subject-matter experts, industry norms, and even weather patterns. When the algorithms detect a probable breakdown, they automatically send the locomotive to a repair facility.

By the time the locomotive arrives, mechanics need not run diagnostics. They pick up an iPad, and in a few minutes, the algorithm reports exactly the source of the imminent breakdown, as well as the locomotive's history and past operating patterns. The mechanics' job is reinvented to do what they do best: instead of waiting for a breakdown and then running diagnostics, their job is now to fix problems before the breakdown, informed by the algorithm so they can use their best experience, judgment, and skill. The mechanics' job is now also to train the automation. Each of their decisions and actions become data that feeds back into the software, constantly improving future predictions.

Source: Brad Keywell, "The Fourth Industrial Revolution Is About Empowering People, Not the Rise of Machines," World Economic Forum, June 14, 2017, https://www.weforum.org/agenda/2017/06/the-fourth-industrial-revolution-is-about-people-not-just-machines; and "About Uptake," Uptake website, https://www.uptake.com/about.

Baxter is a cobot that performs a wide range of assembly-line tasks, including things like line loading, machine tending, packaging, and material handling. Strategically, organizations acquire and deploy Baxter cobots to achieve these strategic goals:

- Safety: Baxter operates safely near humans, without needing a cage, saving money and floor space.

- Trainability: Baxter learns by watching the movements of human workers, reducing or eliminating the time and cost of traditional programming.

- Redeployability and flexibility: Baxter can execute a range of tasks and, because it is trainable, can be repurposed quickly to other tasks.

- Easy integration: Baxter connects with other automation on the line, often without any third-party integration programming or design.

- Compatibility: Baxter's arms move like human arms, so assembly lines built for humans don't have to be reconfigured to work with it.

Baxter isn't the only social robot design. Social robots increasingly come in the form of drones that fly or swim, anthropoid robots that walk, and swarm robots that roll. Traditional robots were mostly limited to very routine and repetitive tasks, but social robots now automate both routine and nonroutine tasks. Freed from the assembly line, such robots can collaborate with humans in ways that were unthinkable before.

Swarming robots are reinventing warehouse operations and shipping at Deutsche Post AG's DHL facility in Memphis, Tennessee, for the provider Quiet Logistics as it fulfills online orders for retailers like Bonobos and Inditex SA Zara.[4] The strategic goals behind these applications of cobots is to reduce the cost of million-dollar fixed conveyor belts and warehouse transport systems. Cobots cost much less—$30,000 to $40,000.

Farmers and Allstate Insurance had the strategic goal to use automation to speed up their response to Hurricane

Harvey victims.[5] They used drones to reinvent the work of claims analysis and payment. Collaborative drones worked beside human claims adjusters to assess property damage. The drones accessed areas that humans could never reach or that were very dangerous. The drones gathered data and took pictures of the damage and sent it to a database. Claims adjusters no longer did the dangerous work of accessing damage sites and gathering data. Instead, they analyzed the database produced by their drone collaborators and made faster claims decisions. Farmers reports that the reinvented job combining a drone and a human claims adjuster can process three houses in an hour. Previously, human adjusters took a full day to process the same three houses. (See the sidebar "How Automation Is Evolving.")

Convergence: Three Categories of Automation Reinvent Oncology Surgery

We've shown how to reinvent jobs using each category of automation. However, convergence means that all three categories of automation work simultaneously. Moreover, work automation seldom affects only a single job. Reinventing one job reveals opportunities and requirements to reinvent related jobs. So, optimizing work automation requires considering all the automation categories and reinventing multiple jobs, reshaping the work of entire teams.

An oncology surgical team provides an example of how work automation is the convergence of multiple automation categories and the reinvention of multiple jobs. The compelling strategic goals that prompt hospitals to pursue surgi-

How Automation Is Evolving

Robots are typically programmed by coding. This places a heavy premium on time and coding expertise. Advances in machine learning enable us to write a piece of code once so the robot is imbued with the ability to learn; we can teach new skills by providing new data. Researchers at UC Berkeley have developed the capability to train robots with new skills in minutes using virtual reality (VR) headsets. Instead of having an expert coder spend weeks training a robot to perform a specific task, Pieter Abbeel and his students Peter Chen, Rocky Duan, and Tianhao Zhang have developed a solution that enables robots to learn by mimicking the actions of skilled craftspeople and factory workers using commodity VR equipment. Instead of a robot taking weeks to learn a new skill, it can learn it in a day using this approach. Says Duan, "When we perform a task, we do not solve complex differential equations in our head. Instead, through interactions with the physical world, we acquire rich intuitions about how to move our body, which would be otherwise impossible to represent using computer code."

These rapid advances in deep reinforcement learning and deep imitation learning will transform how robots are trained and retrained quickly and seamlessly. Think of the speed with which we could deploy robotics in manufacturing plants with these techniques. They address two of the biggest obstacles with automation: the need for significant volumes of clean data and the need for expert coders. Now anyone will be able to teach a robot.

Source: Robert Sanders, "Berkeley Startup to Train Robots Like Puppets," *Berkeley News*, November 7, 2017, http://news.berkeley .edu/2017/11/07/berkeley-startup-to-train-robots-like-puppets/.

cal automation are faster patient recovery, shorter and less expensive hospital stays, fewer diagnostic and surgical mistakes, and being on the cutting edge. Successful execution, however, requires optimizing work automation and reinventing jobs.

A recent article captured the enticing imagery of robotic surgery that often captures the imagination of patients and doctors, and motivates hospital leaders to spend millions:

> Wrapped in plastic sleeves that cover its central boom and sprawling white arms is Intuitive Surgical's da Vinci Xi robotic surgery system. It's hard to tell who's in charge. The instruments inside the patient include three separate, interchangeable components that can slice, shift, grasp, cauterize, or otherwise manipulate human tissue, as well as a movable high-definition camera that illuminates the body's internal landscape in stunning 3D clarity. That's a visual advantage that Sullivan says has revolutionized how doctors perform minimally invasive surgery—the kind that doesn't require chopping someone open to remove a body part or collect samples.
>
> Sullivan makes his way to a console on the left side of the OR, where he takes a seat in front of a viewfinder that looks like it belongs in a futuristic video game arcade. He places his middle fingers and thumbs into two pairs of rings on two movable arms. At the console's floor are foot pedals, which function like a clutch in a manual car. With his fingers and feet, Sullivan will navigate the four instruments now inside the patient's body—alternating between the pincer-laden surgical extensions and a 3D endoscopic camera.[6]

The multimillion-dollar investments in robots, technology, and AI will pay off only if leaders reinvent the work. The oncologist's job typically entails the following activities or tasks:

- Reviewing patient information

- Diagnosing cancer

- Evaluating and choosing treatments

- Executing the selected treatment or surgery

- Coordinating treatment with the oncology team

- Conducting postsurgery monitoring, care, and counseling

RPA, cognitive automation, and social robotics and transform each of these tasks.

Reviewing Patient Information

RPA can integrate the diverse array of information about the patient that resides in different information systems. It integrates the patient's biomarkers, medical history, lifestyle, previous treatments, and so on, creating a comprehensive view of the patient that was previously impossible. New patient information is integrated as soon as it is generated, transforming static information into a dynamic evolving snapshot of the patient.

Diagnosing Cancer

There is no intelligence associated with RPA. Adding cognitive automation with natural language processing, the automated system can now read this evolving data. It can compare each patient to thousands of other patients and score the patient's cancer risk.

IBM Watson for Oncology (WFO) is a cognitive automation platform that achieved 90 percent success in diagnosing lung cancer.[7] Human oncologists average about 50 percent success. Watson ingests over 600,000 items of medical evidence, reads over 2 million pages from medical journals, and searches up to 1.5 million patient records. Its knowledge far exceeds anything humanly possible. Memorial Sloan Kettering Cancer Center estimates that for human doctors, trial-based evidence makes up only 20 percent of the knowledge they use to diagnose patients and choose treatments. A human doctor would need to spend at least 160 hours a week reading journals, just to be aware of new medical knowledge as it's published. WFO can much more quickly and accurately assimilate the vast amounts of new evidence added to the global cancer database and update its algorithms accordingly.

Evaluating and Choosing Treatments

Oncologists should evaluate and choose cancer treatments using the most recent practice and evidence-based guidelines. Can automation match humans in choosing treatments? WFO demonstrated very similar recommendations to a panel of oncologists in a double-blind study of lung, breast, and colorectal cancer.[8] How? WFO extracts and assesses large amounts of structured and unstructured data from medical records, using natural language processing and machine learning to evaluate and choose among cancer treatment options. Approximately

90 percent of WFO's recommendations agreed with those of a tumor board of fifteen oncologists. At first, it took the oncologists an average of twenty minutes to capture and analyze the data and produce recommendations, although they improved to twelve minutes with practice. WFO took forty seconds.

As WFO improves its capabilities, the job of diagnostic oncologist is reinvented. WFO analyzes typical and common cases, allowing human oncologists to focus on the unusual or difficult cases.

Executing the Selected Treatment or Surgery

The earlier vignette about Intuitive Surgical's da Vinci Xi showed how cutting-edge collaborative robotics can actually enhance the act of surgery. Still, humans do most surgery, and most of the advances have been focused on making surgery as minimally invasive as possible. When performing robotic surgery with the da Vinci Xi—the world's most advanced surgical robot—the surgeon controls miniaturized instruments that are mounted on three separate robotic arms, allowing the surgeon maximum range of motion and precision. The da Vinci's fourth arm contains a magnified, high-definition 3-D camera that guides the surgeon during the procedure. In other words, the machine has no intelligence beyond that of its operator. It does not meet the previously defined criteria of social robotics because it has no AI or sensors.

AI is being incorporated into surgical procedures with technology like the Smart Tissue Autonomous Robot (STAR).[9] It uses its own vision, tools, and intelligence to execute surgical procedures. STAR actually exceeded the performance of human surgeons. Researchers programmed STAR to do intestinal anastomosis, in which a surgically cut portion of intestine is stitched back together. In only 40 percent of the trials did human surgeons need to intervene with guidance.

This is a good example of optimizing the human-automation combination. The researchers concluded that the 40 percent of trials requiring human assistance offer clues to designing a new job that involves shared human-machine collaboration in operating rooms. The job of surgeons would now be to supervise procedures and hand over the correct tasks to the robot. In the new job, the automation executes and learns more routine or tedious tasks, leaving the human to focus on the more complex and unusual ones. Automation augments human capability.

Choosing nonsurgical or postoperative procedures is equally as important as choosing and executing surgical procedures. What role can automation play in these procedures? Patients with similar cancers can respond completely differently to the same treatment. These differences can often be predicted based on patient genetics. More precise and personalized treatment requires identifying which genetic factors predict remission or resistance. Does automation have a role here? A team of researchers fed AI the genetic data of cells and tumor tissue from breast cancer patients. The AI algorithms predicted that 84 percent of the patients would go into remission using the drug Paclitaxel. The genetic signature of the drug gemcitabine was able to predict remission using preserved tumor tissue with 62 percent to 71 percent accuracy.[10]

Coordinating Treatment with the Team

The work of coordinating care between the various members of the oncology team is vitally important to the care of the patient. While IT can help ensure consistent access to information and facilitate interaction between the oncologist and the other roles responsible for treating a patient and RPA can replace many activities associated with integrating data from multiple systems, the true value in this particular activity is the result of personal interactions between the various participants. In other words, the work of coordination shifts from one of gathering, reviewing, ingesting, and discussing data to one of insight-driven collaboration where the work is about brainstorming, exploring, and asking "what if?" Cognitive automation can augment such collaboration by applying intelligence to the data about the patient and helping each stakeholder better understand the unique implications of their work for each other and the patient. AI algorithms are able to run countless simulations from different combinations of activities to project and predict various outcomes that may result. This intelligence can then help shift and optimize the behaviors of the various team members toward those that actually can yield a better outcome for the patient.

Conducting Postsurgery Monitoring, Care, and Counseling

A great deal of postsurgical care requires tasks using the sort of empathy and emotion that no machine can do, but cognitive automation still plays a significant role. Cognitive automation, supported by RPA-created data, gathers and analyzes patient data. Caregivers can use these insights to know how different treatments work on patients with certain genetic makeups. The caregiving staff and physicians can now deliver personalized care that increases rates of patient recovery and reduces complications. Automating the routine tasks of an oncologist's job reinvents it to focus on the empathetic and emotional tasks that humans do best, and are vitally important in patient recovery. Humans assisted by the AI insights also are more precise in prescribing drug treatments.

Table 3-3 summarizes our description. Notice how several types of automation converge on the work of cancer treatment. Automation reinvents individual jobs, but also creates opportunities to systematically reinvent the relationships among jobs. Automation both replaces and augments the tasks of oncology and cancer treatment. The goal shifts from

TABLE 3-3

How automation converges to reinvent cancer treatment

Activity	Role of automation	Type of automation
Reviewing patient information	Replaces	RPA and cognitive automation
Cancer diagnosis	Augments	Cognitive automation
Evaluating and choosing treatments	Augments	Cognitive automation
Executing the selected treatment	Augments	Social robotics
Coordinating treatment with the team	Augments	Cognitive automation
Conducting post-treatment care and counseling	Creates	Cognitive automation

executing a process or subprocess to addressing the original strategic goals.

As the half-life of skills continues to shrink, the growing premium on reskilling is causing many organizations to rethink the risks associated with full-time employment in order to reduce the risk of obsolescence. The different variations of work-task automation like the ones here can deliver viable solutions to all of the concerns we've discussed. Selecting the right technology for automating work tasks and improving performance is therefore critical for business, as is the alignment of the selected technology with a comprehensive future of work strategy. Recognizing how technology and AI can transform the performance and value equation provides a significant competitive advantage. Successful leaders will be able to translate the evolving pivot points in their business models into specific implications for work, looking beyond jobs, and to understand the transformative role AI can play in redefining the performance curve for the work of the future.

We will now combine deconstruction, ROIP, and automation into a playbook for optimizing work.

-- -- -- -- --

FOUR

Optimize Work

What Does the Right Human-Automation Combination Look Like?

We've described how to deconstruct work into its elements (chapter 1), how to apply the concept of ROIP of deconstructed work elements (chapter 2), and understand how work-automation applications combine work element characteristics (repetitiveness, independence, and physicality) with the role of automation (replace, augment, or create) and the type of automation (RPA, cognitive, social) (chapter 3).

Here, we combine those threads to understand how ROIP and the characteristics of work elements help you determine the optimum role of automation (does automation substitute, augment, or create new work?), the optimum type of automation, and the nature of the payoff to work automation.

Let's go back to the ATM.

Beyond ATMs Replacing Tellers: Optimal Solutions to Bank Work Automation

When you begin thinking about ATMs, your initial question, "How many humans can we replace with automation?" is naive. Your technology analysts' calculations of the cost savings from replacing human bank tellers with ATMs is well intentioned but actually misdirected. That original question can only lead to the conclusion that the answer is neither yes nor no, and your debate could go on forever, illuminating very little.

You may think that replacing tellers with machines is the question, but when you deconstruct and add ROIP, you realize that some tasks have very different payoffs from changes in work performance or automation and that a teller machine is not the answer to all of them. For some tasks, the payoff comes from reducing mistakes, while for others, it is from reducing variation where it doesn't add value to the customer experience. For yet others, the payoff comes from incrementally adding to productivity and, in yet others, exponentially creating new work or enhancing performance. Then, when you look at the work characteristics, you realize that tellers' work can be rated as more or less repetitive (the more repetitive, suggesting that RPA or simple algorithms might work), more or less independent (more independent, suggesting that you don't need sensors or social robotics to enhance human interactions), and more or less physical (more physical, suggesting the answer will include physical automation like social robotics versus cognitive automation). After you have mapped the work characteristics along these lines, you realize that it would be silly to use teller machines to counsel customers but logical to use them to receive or give cash and count it (i.e., automate repetitive work). You realize that a lot of the payoff of teller machines is in reducing mistakes or just standardizing ways of doing things, which have a certain payoff but will not revolutionize the work or services.

You also realize that some elements of tellers' work, things that are more cognitive and contribute to greater performance value (high ROIP), could be automated, but you need something different from an ATM. You need cognitive automation that enhances the quality of interactions between teller and customer. You also realize that applying automation to some work elements (counting, receiving, and giving cash) will fully substitute automation for human tellers, while for other work elements, automation will create new work that could not be done without automation (accessing the complete customer record of bank services and history).

Fast forward now to today's reality in the ATM story. The actual path of bank-teller work automation can only be seen clearly with the four-step framework that can examine the cost, risk, and quality implications of different types of automation, applied to different work tasks, each with a different ROIP and cost profile.

Bank tellers and their leaders now have immensely more sophisticated automation options than the simple ATMs of the 1970s, and those options are evolving quickly. Some of the cognitive tasks are now done by algorithms or AI that informs and supports the human worker. Still other tasks (such as collaborating with designers and process leaders) are likely to remain in the domain of humans for a long time. The introduction of automation creates new work for humans. We now see the prospect for automation of even more routine processes (e.g., ATMs now use optical sensors and AI to accept deposits of checks and to pay property tax bills) and human tellers providing a much higher-touch service. Yes, this is more complex than simply tabulating how many human workers automation will replace, but it is also much more precise and actionable.

Putting It All Together

We introduced the ROIP curves for the pilot and flight attendant roles in chapter 2. Let's return to those two roles and add our three types of automation to ROIP to show how automation and ROIP combine to reveal new insights. We pose the question, "Can RPA, cognitive automation, and social robotics replace or augment the work of the human being, and what is the payoff?"

Let's take a look at the right side (the steep, upward-sloping portion) of a flight attendant's ROIP curve (see figure 2-3). How might cognitive automation augment human capabilities in delivering the optimal customer experience to shift performance to the far right and up the steep part of the curve? Augmented reality powered by cognitive computing can deliver an unprecedented level of insight into the unique needs of individual passengers. Such automation would augment the customer-care work elements of flight attendants and enable them to unleash their discretionary effort to deliver delivering highly personalized service.

Now, consider how social robotics can affect the left side of the curve (negative-value ROIP) for a pilot. Without automation, the way to reduce mistakes and get performance to be "at standard" involved investing in things like long careers to build proficiency through intensive simulation-driven training to prepare human pilots for any eventuality. Social robotics (the combination of AI with sensors and existing cockpit hardware) can substitute for pilots in all tasks involving routine navigation and even takeoffs and landings, significantly reducing the likelihood of error.

Also consider our example of cancer treatment. Table 4-1 shows how all of the elements of our framework are reflected in the work-automation opportunities for oncologists.

TABLE 4-1

The four-step framework applied to cancer treatment

Activity	Return on improved performance	CHARACTERISTICS OF THE WORK ELEMENT			Role of automation	Type of automation
		Repetitive vs. variable	Independent vs. interactive	Physical vs. mental		
Reviewing patient information	Many different ways, same value	Repetitive	Independent	Mental	Replaces	RPA and cognitive automation
Cancer diagnosis	Avoid mistakes	Repetitive	Independent	Mental	Augments	Cognitive automation
Evaluating and choosing treatments	Better performance produces incrementally more value	Variable	Independent	Mental	Augments	Cognitive automation
Executing the selected treatment	Better performance produces incrementally more value	Variable	Interactive	Physical	Augments	Social robotics
Coordinating treatment with the team	Better performance produces exponentially more value	Variable	Interactive	Mental	Augments	Cognitive automation
Conducting post-treatment care and counseling	Better performance produces exponentially more value	Variable	Interactive	Mental	Creates	Cognitive automation

In summary, the first three steps, deconstructing jobs, describing the ROIP of work tasks, and identifying automation options, come together in the fourth step: optimizing the combination of work and automation.

Another way to see this process is through a series of questions:

1. What are the elemental tasks within the jobs?

2. What are the characteristics of the work?

   a. Repetitive versus variable

   b. Independent versus interactive

   c. Physical versus mental

3. What is the ROIP of the work?

   a. Reduce mistakes (negative ROIP)

   b. Reduce variance (constant ROIP)

   c. Incrementally improve value (incremental ROIP)

   d. Exponentially improve value (exponential ROIP)

4. Does automation substitute for the human, augment the human, or create new work?

5. What are the available types of automation (RPA, cognitive automation, or social or collaborative robotics)?

6. What is the optimal way to combine human and automated work across jobs and processes?

Figure 4-1 shows our four-step framework graphically. This framework can help you navigate the often-daunting tasks of taking high-level strategic automation goals and more clearly

FIGURE 4-1

Step 4: A framework to optimize work: The right combination of deconstruction, ROIP, and automation

Step 1: Deconstruct the work

Repetitive ← → Variable

Independent ← → Interactive

Physical ← → Mental

Step 2: Return on improved performance

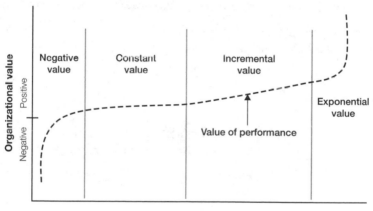

Step 3: Automation type

Process automation
Cognitive automation
Social robotic

Step 4: Automation role

Substitute
Augment
Transform

Examples of optimized work

- RPA substitutes for repetitive, independent, mental work to reduce mistakes
- Social robotics substitutes for repetitive, Independent, physical work to reduce variance
- Cognitive automation augments variable, interactive, mental work to incrementally improve productivity
- Cognitive automation augments variable, interactive mental work to exponentially improve performance
- Social robotics creates new variable, interactive, physical work to exponentially improve performance

identifying how to reinvent jobs and reconfigure work and your organization to meet those goals. One of the best ways to see the framework in action is through examples.

In the next sections, we describe prominent examples that show how the elements of our framework come together to explain work-automation optimization. These examples are not comprehensive, and they do not imply that any one solution always works best. Rather, they show how to use the framework to explain and understand evolving work-automation solutions.

Obviously, when you combine the different elements of each of the four steps of our framework, you can envision many potential work-automation options. In the appendix, we provide a grid describing a comprehensive list of the feasible combinations and show how each example fits that list.

Repetitive, Independent, Physical Work with Negative-Value ROIP: Substitution with Social Robotics

Manufacturers once would have had a whole department of humans to perform inspections of manufactured parts. Here's how Compass Automation, a maker of inspection robots, describes the process:

> This machining center includes an out-feed conveyor, which delivers completed parts to the automated system. Once a completed part enters the system an in-feed nest accepts parts from the conveyor, utilizing a proximity sensor to alert the robot of part presence in the nest. The LR Mate 200iC robot moves the part to a blow-off

station, where excess cutting fluid is removed from the part. The robot then moves the part to the New Vista Thread Verification Unit for inspection; this unit tests the pitch and depth of the part's internal threads.

Next, the LR Mate 200iC robot moves the part to the vision inspection station, where a custom-designed inspection system measures several geometric features on the part, including the part's height and diameter.

Finally, the robot moves the part to an out-feed conveyor for post-processing and packaging. This Compass Automation system, along with the FANUC LR Mate 200iC made it possible for the end user to meet the standards set by their customer for complete inspection of the part at the required production rate. Compass Automation fully models all custom-designed systems prior to beginning the build process.[1]

Inspecting parts has a classic negative-value ROIP. The value is in catching below-standard parts. There is no incremental value from exceeding that standard (e.g., looking for anomalies beyond the standard defects), but there is substantial cost in missing a part that is below standard. Substituting human inspectors with a combination of visual sensors, AI, and precision gauges allows a social robot to move around the factory floor to inspect parts with greater efficiency and fewer errors. The automated solution reduces negative ROIP because it works at the same pace as a human inspector, but with far fewer errors.

This example also applies to work with a significant risk of worker injury or environmental harm. Substituting automation for human workers can both reduce errors and avoid human injuries or death. These are perhaps the most extreme versions of negative-value ROIP. For example, Rio Tinto has deployed automated haul trucks and drilling machines at its mines in Pilbara, Australia. This achieved lower energy consumption and better employee safety.

Repetitive, Interactive, Physical Work with Incremental-Value ROIP: Substitution with Social Robotics

On oil rigs, traditionally the work of "pipe running" (i.e., attaching sequential sections of piping to extend the drill bit to reach oil reservoirs deep underground) is a manual process involving interactions between several workers, with a driller supervising from a control room. Today, automated rigs (e.g., iRig by Nabors Industries) feature automated tubular running services and equipment that eliminate the need for manual labor on the rig floor. The driller in the control room now directs and supervises the iRig.

This automation improves performance (moving upward and to the right on the incremental-value ROIP curve), as a result of faster work and reduced accidents.

Variable, Independent, Physical Work with Exponential-Value ROIP: Augmentation with Social Robotics

How do you quickly detect methane leaks around gas and oil production sites? Human workers usually carry infrared cameras into potentially contaminated fields, walking through the field until they see the infrared image of a leak. It's a slow, arduous process. Even the best human worker can identify only the presence or absence of a leak, but not its size or extent.

Pacific Gas & Electric (PG&E) is testing aerial drone robots to augment the human workers.[2] These drones were first used to detect methane on Mars. Now, they can cruise neighborhoods on Earth, searching for potentially dangerous methane leaks. PG&E worked with engineers at NASA's Jet Propulsion Laboratory and mounted an open path laser spectrometer sensor on a simple off-the-shelf drone. The technology detects leaks faster and more accurately. That creates a payoff, moving up and to the right on the incremental-ROIP curve. In addition, the drone can scan hard-to-reach places where humans cannot go and is thousands of times more sensitive than the handheld technology that humans use. As one manager said, "A cow will generate more methane than the leaks that we are testing here today." The job of inspection is reinvented. Now one worker pilots the drone, while another monitors the readings on a laptop.

Here, social robotics (drones flying among humans) augments the work of the human. First, it improves performance along the incremental-ROIP curve (detecting the existence of leaks faster and more accurately). Second, it makes the human workers exponentially more valuable than they ever could be without automation. This achieves exponential ROIP, shifting performance onto a new steeper ROIP curve (measuring formerly undetectable leaks in places humans cannot access, and measuring both the presence and severity of the leak).

Many related jobs have been reinvented. The drone now does the tasks of gaining access and doing the measurement. The new task of piloting the drone is created. Human inspectors still do the tasks of choosing the site and interpreting the readings. Humans do the tasks of analysis, but now with immensely more precise and complete data. The drone substitutes for humans in some tasks, augments the human in others, and creates yet other tasks, generally shifting human engineers to focus on those with exponential ROIP.

Variable, Interactive, Physical Work with Exponential-Value ROIP: Augmentation with Social Robotics

The Smart Tissue Autonomous Robot (STAR) is an example of this combination (see chapter 3). A social robot with precise sensors and AI takes on the physical tasks of surgery, interacting with the surgeon and the patient, and tackling situations that can vary across patients and surgical conditions. This exponentially increases the value of the human surgeon's performance. Human surgeons can now assign STAR to the more routine or tedious tasks that a robot does better. Every surgeon becomes safer and less risk-prone in these tasks. In addition, STAR augments the surgeon in more complex and judgmental tasks, allowing formerly average surgeons to achieve results comparable to the best surgeons. This creates a new exponential performance ROIP curve.

Repetitive, Independent, Mental Work with Negative-Value ROIP and Incremental-Value ROIP: Substitution with RPA

Financial services companies face increasing costs for compliance. According to the Institute for International Finance, compliance activity can cost about $1 billion a year for a financial institution. Much of this cost is attributable to labor with significant responsibility falling to a credit analyst. That job includes tasks that improve compliance with requirements such as "know your customer." This regulatory requirement dictates that banks demonstrate that they know who borrows money from them or buys their products. The bank must show that they know how the customer uses those funds, for example, to avoid lending to a drug dealer or money launderer. When someone applies for a mortgage, the tasks of the analyst include gathering data from the bank's databases, pulling a credit history from rating agencies, analyzing the borrower's tax records, and scanning social media for indications of borrower risk. This work is extremely time consuming and error prone, resulting in both incorrect decisions and potential violations of state and federal laws.

Now, RPA can perform many of these tasks. RPA gathers data from various sources, integrates the data, and presents a complete picture of the customer with far fewer errors at a fraction of the cost and time. McKinsey & Company estimates that such automation will generate a return on investment of between 30 percent and 200 percent in just the first year.[3] Moreover, reduced errors means fewer defaults, higher customer satisfaction, and lower dropout rates from things like loan applications. These strategic goals can only be accomplished by reinventing the credit analyst job and transforming the approval process. As RPA substitutes for certain human tasks, credit analysts can devote more time to client support (e.g., explaining why a mortgage was denied and offering suggestions for clients to improve their credit rating, and handling unusual exceptions. Here, the job has been reinvented by deconstructing it to separate the tasks of gathering data and evaluating client risk using fixed rules from the tasks of supporting clients and handling exceptions. Automation substitutes for human workers on the former tasks, freeing human workers to create greater value on the latter tasks.

RPA also improves scalability. A spike in mortgage applications due to a cut in interest rates might traditionally require finding forty extra workers on the weekend. With RPA, you can add robots to do the work and then decommission them on Monday.

Optimizing work automation by substituting RPA for human workers improves negative-value ROIP by reducing errors. Shifting human workers to client contact and analyzing unusual cases moves them to tasks that have incremental-value ROIP, where their performance improvements add more value.

Repetitive, Independent, Mental Work with Incremental-Value ROIP: Substitution or Augmentation with Cognitive Automation

Recruiting is an extremely time-consuming activity for every organization. Recruiters expend significant time advertising positions, sifting through candidate profiles, assessing their skills and fit, and then scheduling interviews. This type of work automation has the same work characteristics: as the prior two (repetitive, independent, and mental), but here we shift performance to the right on an incremental-ROIP curve; the automation-work optimization augments human workers versus substituting their work. This example also uses cognitive automation, not RPA.

Cognitive automation can augment some of the recruiter's work tasks. Unilever uses AI to source candidates by placing ads on online platforms like Facebook, WayUp, and Muse. Applicants click on the ads and apply through their LinkedIn profile. Unilever receives hundreds of thousands of such applicants. Traditionally, human recruiters would read these applications, sorting them by whether they were clearly qualified or unqualified. Unilever dealt with this vastly increased volume of applicants by building an algorithm that sifts through applicant qualifications, separating unqualified from qualified applicants. Those qualified complete an automated series of online games, tests, and video recordings. Finally, the AI system organizes the resulting data and scores each applicant using programmed rules, identifying who should receive an in-person interview. Unilever says that it gives an offer to 80 percent of applicants granted an in-person interview using this approach, up from 63 percent before using AI.[4]

Unilever's method for using AI is similar to how the financial services industry uses it for credit analysts, as a sort of "audit" of applicants. AI filtration systems like this eliminate humans from doing repetitive recruitment tasks, which reduces labor costs and human error.

The automation also deconstructs the recruiter's job so that the repetitive tasks are separate from the more human work of actual applicant interviews. Humans now focus on these more high-value tasks, after applicant skills have been assessed quickly and without bias. In addition, humans can now focus on delivering concierge-level service for their highest-profile candidates, eliminating many of the stresses associated with changing jobs. In short, by substituting automation for humans in the repetitive tasks, recruiters are available for the nonrepetitive and interactive tasks of helping candidates navigate their way through the recruiting and on-boarding process, serving as a one-stop shop for all their cares and concerns.

Sometimes, cognitive automation augments human workers by interacting with them. Kensho Technologies' investment analytics system allows financial investment managers to ask questions in plain English and get answers in seconds. They might ask, "What sectors and industries perform best three months before and after a rate hike?" AI becomes their natural-language conversational adviser. Every human analyst doing the tasks of exploring different future scenarios now produces far greater value more quickly.

Both examples focus on tasks that include repetitive, independent, mental work with incremental ROIP. Cognitive automation substitutes or augments human work, moving performance to the right and up the slope of the incremental-ROIP curve by enhancing performance in assessing a candidate or investment. Automation analyzes more information faster and with higher quality and less bias. As we have seen, this substitution also allows the organization to shift human workers to higher-value-added tasks.

Repetitive, Interactive, Mental Work with Exponential-Value ROIP: Augmentation with Cognitive Automation

How do you maintain fleets of aircraft, vehicles, and windmills? Traditionally, an army of GE's skilled and experienced technicians would visit customers' locations to do maintenance. The job included the task of deciding what maintenance was needed, using rules of thumb based on factors like how long the machine had been running, the load and environmental conditions, and so on. The technicians' job also included the task of sharing their experiences (or the best practices they discovered) by sending messages to electronic bulletin boards or processing centers, accessible to other technician teams. The work is repetitive because they execute the same tasks in a specific sequence. It is interactive because technicians on each on-site team must collaborate, and separate teams collaborate on the common learning platform. It is mental because it involves deciding what information to gather and then analyzing the information to choose the best maintenance approach. Of course, the job of a maintenance technician also includes physical tasks to execute the activity, but here we focus on the data-gathering, analysis, and diagnosis tasks. The traditional approach using human technicians produced significant equipment downtime, because customers waited for technician teams to be scheduled or teams faced new situations that required learning time, or because the central information platform waited for updated information on best practices from the field teams.

GE's strategic goals were to drastically reduce downtime, perform maintenance only when needed, and offer more complete and customized data and best practices to human technicians in the field. GE did this by reinventing the technician job using AI, specifically machine learning, to leverage the power of sensors, big data, and IoT.[5] GE created a "digital twin," a digital replica of equipment such as a jet engine, gas turbine, or windmill. Sensors in the actual equipment gather data on the machine's attributes and operating environment (heat, vibration, noise, etc.), and the data is organized into the simulated digital twin that now replicates the physical machine's performance. GE can then program the digital twin to simulate different scenarios (loads, durations, environmental conditions, etc.). Using the data from the simulation, AI can diagnose faults and predict maintenance needs. AI can then create maintenance schedules and send alerts and recommended best practices to GE's human technicians in the field. Multiple digital twins can even be linked to mimic aggregations of actual equipment (i.e., an entire factory) or vehicles (fleets of aircraft or trucks), allowing analysis of not only the performance of each individual machine, but the collective and integrated behavior of the network of machines.

Data from thousands of actual machines flows perpetually and constantly into the digital-twin models. Because the variables affecting machine performance and optimum maintenance change with time and circumstances, optimal maintenance is not simply a matter of finding one formula and following it consistently. With human technicians, the best available solution was often to take one consistent approach, because it was not possible to customize the approach to each situation. With automation, the optimization algorithms and approaches can be constantly updated with the flow of additional data. This machine learning allows the technology to learn from new data and modify predictive models over time, identifying anomalies and trends and understanding patterns. Machine learning can identify efficiencies in one machine or situation and apply it as a best practice for others. In 2017, GE had about 750,000 digital twins and was adding more.

In this situation, machine learning requires a combination of sensors, IoT, big data, and Web 2.0. A system without machine learning must rely on what a single customer can observe, or what individual technician teams can learn and convey to other teams. Optimizing the combination of humans and machine learning allows GE to leverage vastly more data, analytics, and learning across all the organizations that use its products. There is also a network effect, because the more GE learns, the more customers benefit by choosing GE engines, which grows the network, enhances the learning, and so on.

This has enhanced the performance of technicians on maintenance teams by shifting them to the right along the incremental-value ROIP curve, but it also creates an exponential-value ROIP curve for their performance. That's because machine learning augments the technicians' performance on tasks like choosing maintenance schedules and approaches. Now, when a technical team arrives at the equipment location, they are pre-briefed on the best maintenance approaches for that particular equipment. Those approaches are based not only on the team's experience, but on all the data from all the other similar equipment and on the information from that particular machine's twin. Every technician team is now exponentially more valuable because technicians now are assigned in exponentially more efficient ways, arriving only and always when most needed. They also execute the maintenance activities that are exponentially more correct.

Variable, Independent, Mental Work with Exponential-Value ROIP: Augmentation with Cognitive Automation

Product design and development is variable, independent, mental work. Cognitive automation accelerates product development by augmenting the work of procurement planners, improving their insights.

Coca-Cola Company's Black Book model supports its strategic goal of producing its Simply Orange juice with a consistent taste profile, even as its sources of juice change due to weather and other crop factors.[6] The Black Book model is cognitive automation that uses algorithms to predict weather patterns and expected crop yields. Those results inform the company's ingredient procurement for Simply Orange juice, achieving similar juice taste even with wide variation in the quality and quantity of crops. The automated model supports minute-by-minute updates to procurement plans if weather conditions threaten to damage crops. In the past, the process relied on human planners who simply could not gather and analyze the necessary data quickly enough. There was significant variation in both product quality and quantity due to slow or inappropriate reactions to unforeseen weather conditions. Now, the Black Book model provides human planners with far more precise and rapid recommendations, increasing the value of their performance exponentially.

Stitch Fix, an online fashion retailer, sells fashions that customers don't even know they want. How does it predict customer desires even before they know them? It does this by blending machine intelligence with human intelligence. It reinvented the job of its 3,400 fashion stylists who interact with customers, mostly while working from their homes. Stitch Fix's customers begin shopping by filling out a lengthy, interactive questionnaire, powered by AI. If a customer says she wears medium-sized blouses, the interactive questionnaire is programmed to ask whether they typically fit loosely or tightly. Other questions include: "Does your office require business attire or casual?" "Do you take wardrobe risks?" "Which of these fifteen colors would you wear?" "Do you wear your jeans skinny, straight, or both?" The question-naire data is fed to a database, along with data gathered from the web, such as social media profiles, Pinterest style boards, and so on.

Customers receive periodic orders of clothing chosen for them. They do not see the clothing until they receive it; they pay a $20 "styling fee" to receive each box of clothes, but they can send anything back for free. The boxes of clothes—called a "Fix shipment"—are assembled through the work of human stylists augmented by AI. The AI styling algorithm selects items it calculates that the customer will like. Those selections are sent to the human stylist. The stylist has been matched with a compatible customer, again by an AI algo-rithm. The human stylist then adjusts and streamlines the computer's choices to produce the assortment that is ulti-mately delivered to the customer.

This combination of AI and human fashion selectors reveals nuanced ways to combine human and automated work. Stitch Fix's chief algorithms officer, Eric Colson, says, "It turns out there are things humans can do much better," like curation, seeing things as a cohesive set, improvising, and relating to other human beings.[7]

The job of stylist has been reinvented as an integrated collaboration with cognitive automation. Human stylists do tasks where their performance can add value, like curation, improvising, and relating to others. Cognitive technology takes on tasks that humans do less well, such as gathering and analyzing data, and producing decision guidelines. More important, the results of the cognitive automation make the value of every stylist's performance incrementally better, because they can use the decision rules and results of cogni-tive automation as their starting point.

Variable, Interactive, Mental Work with Exponential-Value ROIP: Augmentation with Cognitive Automation

The earlier examples focused on independent work. Here, we discuss examples that are similar in that they are also mental and variable, but instead are interactive. Again, cognitive automation augments the human worker, but the interactive nature of the work means this happens very differently.

Call-center jobs are very interactive, and not always in a pleasant way. Traditionally, human call-center workers learn whether a customer is angry only after they answer the call. Or, they cannot even tell if a customer is angry because they interact only through text or chat. This means that the initial response to an angry or frustrated customer is often very generic or unemotional. Also, the particular call-center representative who receives the call might not be the best at handling irate or frustrated customers.

Ocado Group, a UK-based grocery delivery service, uses Google's AI tools to analyze language and convert speech into text.[8] These tools identify the irate customers by spotting language patterns in emails and phone calls that Google's research shows are associated with anger, frustration, and irritation. Now, customer-care specialists are notified that the customer is angry before they answer the call. They can respond with the appropriate empathy and emotion, defuse the tension, and perhaps even turn a detractor into a promoter.

Like the GE technicians who have the benefit of the digital-twin analysis, here automation augments the human worker by providing otherwise unavailable data-based insights. In this example of very interactive work, the cognitive automation also improves the interaction between customer and service provider. This makes all customer representatives immediately incrementally more valuable than when they had to guess if customers are angry.

This kind of work-automation optimization also applies to jobs that use research knowledge to deliver creative solutions. A good example is the job of a trial lawyer, which used to involve the task of reading relevant court cases for precedents and judgment patterns. The job can now be reinvented to include asking the AI of IBM's Watson about relevant court cases before developing a defense strategy. Or the job of an architect can now be reinvented to design buildings with the help of AI systems that analyze weather, traffic, demographic and social preferences, and location topography to suggest combinations of building orientation, location of amenities, and so on.

In every example, the reinvented job shifts some tasks to the automation, keeps other tasks with the human, but also augments the human worker to create performance or new work that would not be possible by either automation or the human alone.

Transforming Insurance by Reinventing Entire Job Families or Processes

Our previous examples looked at single jobs or closely related jobs to provide clear illustrations. In reality, optimizing work automation typically transforms many jobs involved in complete processes. Automation simultaneously substitutes, augments, and creates new human work. We will describe the implications for leadership and organization design in later chapters, but now we'll discuss this process transformation using the insurance industry.

The insurance sector is ripe for work-automation optimization. Its global and statewide patchwork of regulation creates extreme inefficiency. Its convoluted distribution models and obscured data insights have long limited its product pricing. For example, auto insurance premiums are mostly determined by a driver's credit score, but that score predicts less than 30 percent of differences in driver risk.

RPA, cognitive automation, and social robotics have converged to transform property and casualty (P&C) insurance to address these broad strategic challenges. Strategic execution requires deep work analysis and clear decisions about how to optimize combinations of human and automated work, and reinvent entire groups of jobs to support that optimization.

Claims Process before Automation

When you are involved in an auto accident, the traditional claims process involves many steps and human workers.[9] First, you call your insurance company. You talk to a customer service representative, who enters details of your accident into a database. That triggers a notification to either send a claims appraiser to your location or give you an address to take your car to. The appraiser evaluates the damage, assesses the cost of parts and labor, and then tells you the insurer's estimated payment to you to repair the damage. You choose from among the approved repair shops and inform the claims appraiser, who notifies the repair shop. The appraiser enters your choice into the system, which triggers the system to send the damage assessment and estimated price to your chosen repair shop. Once repairs begin, the repair technicians might discover additional damage that was overlooked by the field appraiser. That triggers more notifications, a visit or call for the appraiser to reconcile the damage estimates, and more negotiations. Finally, after reaching agreement, the repair shop can complete the repairs and be reimbursed by you or your insurance company. This process can take several days or weeks. It is prone to errors. Moreover, it has little memory, because very little intelligence is generated, even though there are hundreds of these transactions each day.

Claims Process after Optimizing Work Automation

Obviously, there are many ways that automation could enhance this process. It is tempting to concoct new approaches that substitute technology for human workers, insert technology in every transaction to reduce costs and errors, and so on. However, these ideas require precise and careful work design that optimizes the human-automation combinations. That means using automation where it makes sense, avoiding it where it doesn't, and reinventing jobs to reflect the optimal combinations.

What is the new optimized process? The first work-automation step doesn't even involve insurance company jobs. Virtually everyone has a cellphone with a camera, so you can optimize the combination of insurance customers and their cellphones. The policyholders can use their phones to take a picture of the damage. They upload that picture to their insurance company's database. If the damaged vehicle is in a remote or dangerous location, or if the customer is hurt or unable to take a picture, a drone controlled by an insurance company operator can be dispatched to take pictures of the damage.

Once the pictures are in the company's databases, the images are linked with the company's records of the make and model of the vehicle. Cognitive automation then goes to work. These algorithms perpetually learn from the thousands of images of damaged cars uploaded every day and from millions more images already stored in the company's databases. It can analyze the image to determine the damage location, such as the bumper and right rear fender. In fact, it is even possible for your phone to recognize the type and nature of the damage before the images are sent to the insurance company. This doesn't require insurance companies to build image-recognition AI. In 2017, camera chip makers were already adding "smarts" to cameras, allowing the camera to extract information such as the movement of packages or recognizing a person's face.[10] Such technology can actually create automated image-recognition technology that assesses damage more completely and precisely than a human appraiser. The technology can recognize when a damage assessment is unusual or obscure, and forward its assessment to a human appraiser for verification or further analysis. This technology creates new work as human appraisers train the technology.

The next step uses RPA. It finds and connects data on all previous claims involving similar vehicles with similar damage, and related data on parts and labor costs from approved body shops in the area. With no human intervention, RPA and AI can create a detailed appraisal that determines the specific cost of repair, time to completion, and the likelihood of additional repairs being required.

The jobs of appraisers are reinvented. Cellphone cameras and drones substitute for the routine tasks of field appraisal. Cognitive automation now substitutes for the routine tasks of damage estimation. The job of appraiser doesn't go away, but it is reinvented. Now, the appraisers' tasks emphasize reviewing the automated analysis. They can also concentrate more on providing the "human touch" to the involved stakeholders (the policyholder, body shop, other involved drivers, etc.). The appraiser job evolves from executing routine individual transactions to reviewing the output from automation, overseeing multiple stakeholder interactions, and providing concierge-level service.

This process now takes hours, rather than days. The damage analysis from the optimal combination of human and automated work is more precise, and the claim preparation is already based on thousands of similar cases. So, the quote presented to the repair shop is more complete and precise, reducing the chance that overlooked damage will be discovered, or that the quote will omit important parts or labor. However, these attractive strategic outcomes were not achieved by simply substituting algorithms for analysts. They required reinventing several jobs and processes.

So far, our example has focused on making the process more efficient and precise. However, this new optimized process produces an even larger and more strategic benefit by changing the very nature of insurance pricing.

Recall that traditionally insurance was priced by using the policyholder's financial credit score as a proxy for their accident and payment risk. Now, the information from the improved claims process transforms insurance pricing Insurance companies can price their products to account for the historical riskiness of the region where the policyholder typically drives, such as the number of accidents that historically occur on preferred driving routes. It can account for the precise historical cost of parts and labor for the policyholder's particular vehicle and even the policyholder's driving style. Again, this is a game-changing strategic benefit, but it is only possible if companies get the work optimization correct.

The disruptive strategic change doesn't stop there. At the same time that automation improves things like cameras, drones, and algorithms, the very nature of the automobile itself is changing. Driverless vehicles are increasingly widespread. When you combine driverless vehicles with enhanced claims and risk estimation, the very focus of insurance pricing changes. Traditionally, the human driver is the focus of risk estimation (where you drive, how you drive, whether you maintain your vehicle, etc.), and insurance is priced based on the individual.

With driverless technology and enhanced claims and repair analysis, the *vehicle* becomes the risk factor. Now, your insurance company can "microprice," adjusting your premium minute by minute based on the condition of the vehicle, the weather, the location, and the driving style. For example, your insurance premium might spike every time you take over control of the vehicle and drop if you use the vehicle's autonomous capabilities. When AI and sensors can track vehicles continuously, the auto manufacturer can become the insurer.

That's exactly what's happening as Tesla begins to offer vehicle insurance. In Asia, Tesla has partnered with established insurers to offer customized auto insurance that accounts for its vehicles' autopilot safety features. As Tesla advances toward fully autonomous vehicles, the company is in a unique position to actually compete with traditional P&C insurers. Tesla, with more complete data and better analytical technology, can offer lower insurance prices because it has the ability to offer micro-pricing more precisely than those insurers.[11]

The strategic implications stretch even beyond more precise pricing and claims estimation. You have probably guessed that if automation allows for continuous monitoring and analysis of vehicle behavior and performance, it enables a shift from insurance that pays for accidents after they happen to prevent the accidents in the first place. That requires extending process automation beyond claims analysis and pricing.

Willis Towers Watson recently announced a partnership with Roost, a maker of smart-home technologies such as cognitively automated security cameras and thermostats.[12] The partnership aims to bring smart-home device companies and insurance companies together to allow property insurance companies to use AI and the IoT to better mitigate insurance losses for water and fire. Continuous monitoring and analysis of data flowing in from smart-home devices can allow for more precise property insurance pricing and claims analysis, but it can also mitigate insurance claims. An insurance company might offer policyholders a kit with the Roost smart smoke alarms and smart water-leak detectors, and an agreement to send notifications to customers' smartphones when a suspected water leak or fire occurs. This cross-industry integration offers the possibility that insurance carriers can achieve insights from home telematics by capitalizing on the data already being gathered by the telematics companies, rather than making big investments in their own proprietary telematics, and enabling them to more quickly shift from loss reimbursement to risk mitigation.

Imagine how optimizing the human-automation combinations reinvent jobs throughout this process. There will be fewer field adjusters doing dangerous or tedious work, as AI shifts tasks to the customer and drones. Former risk and compliance managers will see routine tasks shift to bots that are less error prone or fraudulent, supported by sophisticated cognitive models that predict quality and compliance. In call centers, managers will shift from overseeing call details to operating sophisticated robo-advisers. Salespeople will spend less time explaining routine features to customers, because customers will use algorithmic bots that offer insurance products precisely tailored to their driving habits, location, and home features. Those algorithms will run on global platforms that link consumer products like phones and thermostats with the databases of retail, financial, and even entertainment and communications services companies.

But, none of this is possible without a clear and precise way to reinvent jobs. That requires a process like the one we have described in this and preceding chapters. (See the sidebar "A Checklist to Help Leaders Implement Automation.")

Now you can see how to answer automation questions in your organization more completely and precisely. You can get beyond simplistic ideas or grand strategic visions that are impossible to execute and instead apply automation to work in a thoughtful and nuanced way that optimizes the value of work and the role of automation.

Optimizing human-automation combinations at the work level is vital. However, your work isn't done there. Even the best work-level automation can fail without a supportive context. Once you optimize work automation, you must consider how that redefines your organization, the meaning of leadership, and your approach to work and careers. The next chapters address these topics.

A Checklist to Help Leaders Implement Automation

As you consider the four steps of our framework, the following checklist may help you get started with planning your automation journey:

1. Identify the opportunity

    *   Is there an opportunity to significantly reduce cost through automation?

    *   Are there new or emerging capabilities that you can develop through automation?

    *   Are you struggling to find talent for critical work areas where deconstruction and automation of certain tasks may make it easier to get the remaining aspects of the work done (either by existing staff, contingent talent, or more easily accessible talent)?

    *   Can you select a pilot location to experiment with automation?

2. Understand the work

    *   What do the characteristics of the work and the ROIP (steps one and two) tell you?

    *   If grouped into a job, can these activities be separated with minimal breakage?

    *   Is there valuable connective tissue between activities that will be critical to maintain? Are there means for doing so that go beyond aggregating them in a job?

    *   What is the expect return from automation (productivity, speed to capability, cost, risk, etc.)

3. Apply automation

- What insight does your analysis of the role and type of automation (steps three and four) yield?

- How will you acquire this automation? Is it available as software as a service (SaaS), as is the case for numerous RPA and cognitive automation solutions? Or, in the case of social robotics, can you lease the equipment from a manufacturer?

- What customization of the automation is required? Who will do this and how?

4. Ensure proper oversight and governance

- Which functions of the organization (HR, IT, procurement) will need to get involved as you apply automation?

- What will be the role of each stakeholder? Who will have primary responsibility for coordinating work in the newly automated workplace?

- How will you train managers and employees to work with your new automation solutions? What new or different skills are required?

- What are the security implications of automation (think potential new cybersecurity vulnerabilities)?

5. Measure your ROI

- How did the actual return from automation compare to the expected return?

- What factors caused the deviation?

- What do you need to change as you move from pilot to full-blown implementation?

PART TWO

Redefining the Organization, Leadership, and Workers

Automation Implications beyond Reinventing Jobs

Recall the conclusions of the recent MIT conference on the future of work: reaping the benefits of automation relies on leaders who optimize the combination of human and automated work and organize to support it.

Part One of the book has shown you how to deconstruct and reinvent jobs to optimize work-automation combinations. Part Two describes how to organize and lead to support and enhance those work-level solutions.

Chapter 5 describes how work automation requires rethinking the very meaning of the organization, including examples that redefine it as a hub of entrepreneurs. You will see how your decisions on optimizing the work can and must be supported by equally sound decisions about organizational factors such as structure, power, culture, accountability, and information.

In chapter 6, we discuss how to reinvent leadership to reflect the constant evolution of work and changing human-automation combinations. Our examples show that leadership will be determined less by fixed institutional structures and become more democratic and more socially determined and more fluid, as work is perpetually upgraded and reconfigured.

These organizational and leadership transformations are built upon work-level optimization, but they are also necessary to maximize the social and economic value of that optimization.

Finally, in chapter 7, we suggest how you can apply the framework to your own work and career. We invite you to reinvent yourself by deconstructing and reconfiguring your work, capabilities, development, and career paths, using the principles of reinventing jobs from part one. Work will constantly evolve faster than ever. So, all workers and leaders must redefine their personal relationships with work as a constant process of reconfiguration, reoptimization, and reinvention.

FIVE

The New Organization

Digital, Agile, Boundaryless, and Work-Centric

Have you ever played Jenga? Players take turns removing one block at a time from a tower constructed of fifty-four blocks. You then place each block you've removed on top of the tower. This creates a progressively taller and more unstable structure. Remember that feeling of removing one of the blocks holding a platform in the multistory tower and seeing the tower start to shake? Will it tip over or will it hold? Organization structures can be like the Jenga tower. You might be lucky, and changing one element, such as reinventing a job through automation, won't affect the overall stability of your organization. More often, reinventing a job has collateral effects on other jobs, the relationships among them, and the communication, authority, and power structures of your organization. Often, if you're not careful, reinventing only one job may make the organization unstable.

How can you identify or predict how job- and work-level changes affect the larger organization? Which reinvented job changes might even improve how your organization functions, and which are like the flaps of the butterfly's wings that cause a tornado?

This chapter examines the organizational implications of optimizing work automation by reinventing jobs. Just as the blocks in the Jenga tower are decoupled from each other, the work elements and tasks that were once contained in a stable organizational structure of jobs will be increasingly decoupled from those former jobs and even from the organization. As we have seen, this unmooring of work tasks offers immense new opportunities to reinvent jobs in new ways. What are often overlooked are the new dilemmas and risks created when you remove the traditional safety net of job descriptions and organization designs based on them. It's the difference between assembling a Jenga tower using a premade framework where you can only assemble blocks in a limited number of ways versus assembling the tower with unlimited freedom in the location and combination of the blocks, and when automation can change the very nature of every block at any time.

The four-step framework in chapters 1 through 4 helps you decide how to deconstruct and reinvent jobs. Those decisions also affect organization-level issues such as culture, diversity, alignment, engagement, authority, and accountability. Those organizational implications are often not obvious if you simply focus on reinventing single jobs. Sometimes, the organization-level implications will amplify the positive effects of work-level decisions. At other times, organization-level implications will resemble the tornado that collapses the Jenga tower. You may even decide to avoid reinventing some jobs, even though it seems logical at the job level. Savvy leaders must avoid being seduced by cost, risk, and productivity benefits from work automation applied to single jobs and consider the organizational implications.

The Outside-In versus Inside-Out Approach

There are two ways to think about the connection between work-level automation decisions and organizational implications.

First, you can work inward starting by understanding how automation might enable the transformation of your industry, strategy, business, and organization. Then, you consider how the work must change to support those organizational and strategic outcomes. A big idea from top officers often prompts this approach. For example, Tony Hsieh, the CEO of Zappos, seems to have had this in mind, as recounted in this excerpt from "The Unorganization," a report by SogetiLabs:[1]

> Tony Hsieh, the CEO of Zappos, sent an extensive email to all employees at the end of April 2015 explaining that they were going to be a company that organizes differently. The world changes quickly and business has become too unpredictable. Hsieh wanted to prepare his organization for this digital era. He wanted to put an end to old-school management, at least the last remains that were still present at Zappos. He said that everybody at Zappos must be adaptive, flexible, inventive and creative. In the email Hsieh

announced that they would continue forward without managers. Self-organization was already a pillar of the culture at Zappos, but Hsieh's ambitions went even further in implementing the so-called "Holacracy," a self-organization methodology introduced in 2007 by the HolacracyOne company.

Executing such a vision requires aligning work and workers in new ways. Will you simply remove the work of the eliminated managers? Or, more likely, will some of that work be decoupled from the former job of "manager" and integrated into other jobs? Success will depend on optimizing work automation by replacing tasks previously done by managers, such as team communication, detecting exceptions that require attention, and analyzing and summarizing data from customers so that teams can quickly respond. The four-step framework for reinventing jobs in chapters 1 through 4 can be applied to issues like these to create the better-optimized new work necessary to support the organizational and strategic vision.

A second way you can connect the reinvention of jobs to your strategic challenges is to work outward by starting with job-level work-automation opportunities that enhance the cost, risk, and productivity value of existing jobs. Then, you must decide how new organization elements must support those newly reinvented jobs, or how negative organization-level implications might suggest reconsidering, slowing, or abandoning those job changes until the organization is ready.

For example, it is quite feasible to automate assembly-line jobs with advanced AI and social robots. However, at the organization level, production innovation ideas come from having humans on the assembly line so they can see ways to make the line more efficient, safe, or customer-friendly, and then communicate those ideas to operations engineers and designers. Remove the humans from the assembly line, and the line itself certainly becomes faster, more reliable, and safer. But, you lose the organization-level value of human assembly workers who find improvement opportunities and communicate those opportunities to production and product design teams.

Let's see how each approach to organization-level optimization works. We will illustrate the outside-in approach with global appliance maker Haier. Then, we illustrate the inside-out approach by extending our earlier example of cancer surgery and treatment.

Working from the Outside In: The Story of Haier

Consider how automation changes the concept of a refrigerator.[2] When sensors and AI can seamlessly and continuously connect it to the cloud, and those connections can, in turn, link to the cloud-based and sensor-powered systems of grocery stores, delivery vehicles, and suppliers (think Whole Foods as part of Amazon.com), the single appliance becomes a hub connected to a vast network of food suppliers. It becomes the entry point for "food as a service" that monitors household inventory levels, orders food, and delivers a fully stocked refrigerator and pantry. Such a transformation is only possible with a combination of technologies such as the IoT, cloud-based storage and services, and powerful cognitive automation and AI. Revenue now comes not only from selling the appliance, but from offering the best user experience in finding, acquiring, storing, and using food, not to mention the value creation and business opportunities arising from data collection, analysis, and interpretation.

How should we design an organization to address the challenges of such transformations? What happens when a manufacturing giant transforms itself to fully take advantage of these technologies?

Haier CEO Zhang Ruimin has become legendary for taking a traditional, hierarchical, global manufacturing organization and transforming it into a platform for serial entrepreneurship, with employees who act as self-governing entrepreneurs. According to an article in MIT's *Sloan Management Review,* Haier has transformed from a manufacturing corporation to a platform providing financing, support, and coordination for a coalition of microenterprises, all focused on building products and services for the "smart home."

Zhang's concept is to turn customers into users who collaboratively improve and develop products by sharing their behaviors and ideas, and to reduce the distance between the organization and these users to zero, so that the organization is constantly co-creating with them. The enabling factors in this transformation are the IoT, a combination of technologies that allows information about the user experience to be constantly gathered, analyzed, and shared with the organization as it designs, develops, executes, and improves products and services that contribute to that user experience. Haier transformed its organization structure from a traditional hierarchy to a platform that encourages employees and partners to join microenterprises operating on the platform. The organizational platform has little hierarchy, but instead operates to provide support and market evaluation for more than two hundred entrepreneurial teams, each equipped with money, technology, logistics, and other support, and rewarded and empowered to act on the information they receive about the user experience in the smart home.

Compensation is determined by how much value you create for the users. As Zhang said, "When employees create value, they get paid. If they don't create measurable value, they don't get paid. Ultimately, if they don't create value, they have to leave." Instead of organizational units, divisions, product lines, and functions, Haier is organized as a goal-based entity, gathering and dispersing resources based on client needs. (更换成 "围绕用户需求，按单聚散的全流程并联组织" 更准确一些.)

As the CEO describes it:

> The process begins with an objective. For instance, someone comes up with an idea for a product targeting a certain niche of the market. And then people from different departments or disciplines—research and development [R&D], sales, manufacturing, marketing—will sit down and analyze its viability across all the relevant dimensions. If they believe it is viable, they will form a community to bring it forward as a new microenterprise. Then they need to attach their plan to their compensation. We call it a predefined value adjustment mechanism, or VAM, which defines what goal the plan has to realize and how the members of the community will be paid if the goal is achieved. This is a signed agreement between Haier and its microenterprises. We also have microenterprises that focus on more cutting-edge projects. These teams may not plan to achieve revenue for a couple of years. Here, we set different targets and schedules. For example, at a certain point of this endeavor, they must be able to attract external venture capital. If they can't achieve the investment by an agreed-upon time, then

they have to let it [the project] go, or we might invite another entrepreneurial team to work on the project.

In our model we have delegated the major powers of corporate executives to the employees—or at least to the microenterprises—including the power of decision-making, the power of selecting and appointing personnel, and the power of financial allocation. Other companies would not do that. They believe that if these powers are delegated, managers will lose control. Our goal is different: We are trying to motivate employees to unleash their potential and realize their own value. We don't want to control them.

The same "close to the user" approach underlies the role of business functions that provide services such as manufacturing. For example, Haier has 108 factories globally, each with multiple production lines. At Haier, each production line functions as its own microenterprise. Zhang describes it this way:

We evaluate the performance of these microenterprises based on cost, delivery and service quality, and market response to the products they make. This evaluation determines how they are qualified to get subsequent orders. Some production lines are able to acquire many orders. Some get fewer—and as a result, employees on those lines are not paid as well. Lines gaining more orders can merge with those having fewer. In this way the production lines are organically connected with the market.

Based on the aforementioned information Haier provided to us, we assessed the Haier case study, its relevance to this book, and the potential implications: the Haier example is a rather radical illustration of the power of technology to disrupt the concept of the organization. It is also an object lesson in how leaders must think about work and automation at several levels. At first, the issues may appear to be merely technological opportunities (AI, cloud storage, sensors, big data, etc.). As we have seen in earlier chapters, optimizing such opportunities requires deep attention to their impact and integration with the work. For example, the IoT allows AI and sensors to be built into refrigerators and other appliances, producing streams of data about operating conditions, but also about the consumer goods that are stored and used.

At a strategy level, the technological capability to link multiple products with Amazon-like services fundamentally shifts the very nature of strategy from making and selling great products to building the connected infrastructure that makes it effortless for users to manage their food acquisition, storage, and preparation.

Consider what happens at the work level in one job of a customer service representative. Such representatives now become AI-enabled. The old job included a work task of interviewing customer callers to discover problems. That work task is now eliminated with AI, augmenting the workers by allowing all representatives to start the call with much deeper knowledge about the customer's problem. Automation also creates new work, because the appliance can actually track food usage and spoilage. Now the customer service representative can coach customers on how to use their personal device to connect the appliance to their Amazon or Alibaba account, so that they can place orders at Whole Foods or other grocery stores automatically. See how the work-automation framework from earlier chapters helps to work outside-in by starting with Zhang's vision of "zero distance to the customer," and reveals the implications at the work level that are keys to its execution.

You can also work outward from these work-level opportunities to see mid-level organization implications. The big vision is "zero distance to the customer" and an organization that is a "hub for data-connected entrepreneurs." Between that vision and the work-design implications for jobs are vital relationships between teams, units, and functions. It is vital to connect the work-automation decisions described in previous chapters (that occur at the level of the work element or job) with the mid-level organizational consequences (that occur at the levels of teams, units, functions, and the organization). This is where issues such as trust, authority, accountability, information sharing, social networks, and cross-unit culture come into play.

For example, if we add the customer-coaching work described earlier to the call-center representative job, those representatives will discover ways to improve the product interface with Amazon and other cloud services. Is that the work of customer service representatives or product designers? Should customer service representatives now also formulate new product design ideas that arise as they coach customers on how to integrate the refrigerator with their Amazon account? How will Haier's talented refrigerator designers react to the prospect that their work tasks will now be automated or transferred to customer service representatives? Maybe this work now becomes part of the refrigerator product design team and gives them the work task of coaching customers, augmented by AI. If Haier leaders rush too quickly to embed customer coaching in the call-center representatives' work, they may inadvertently create conflict or territorial battles between customer service and product design teams, each of whom thought it was their job to tap the power of AI to work with customers to redesign products.

Taking this further, many Haier products, not just refrigerators, can now interface with Amazon. Perhaps the work elements that involve parsing and analyzing how customers use all the products in their home could migrate to a new functional division that never existed before. That division might now have the authority to oversee product design for all Haier products. It would contain work tasks such as listening to customers (or using AI to analyze customer interactions with service representatives) as well as working closely with Amazon or Alibaba engineers to devise creative ways to integrate all Haier products with those services. Again, what began as a work-automation analysis in the job of call-center customer service representatives now reveals questions about creating entirely new divisions, all because deconstructing the job liberated some work tasks to tap the power of AI.

These options require thinking beyond the work of customer service representatives, product designers, or Amazon liaisons. The decisions made at the work-automation level will have a ripple effect on organizational issues such as power, authority, accountability, information sharing, status, and culture. For example, a decision to create a new liaison division devoted to the Amazon or Alibaba relationship, spanning all Haier products, fundamentally changes the power and authority of the former product design group, which now will look to the liaison group for key information about product features and design opportunities. Product design leaders who formerly held the consumer-usage information may well chafe at what appears to be a structure that makes them subservient to this new division that is a liaison to Amazon and Alibaba.

Marketers and salespeople who formerly held great power and authority based on their knowledge about particular products and the channels that sold those products must now evolve to selling based on their expertise in understanding how multiple products connect with each other (such as the refrigerator, personal device, and microwave oven) and channels like Amazon and Alibaba. A leader considering work automation applied to sales and marketing jobs might use the four-step framework in earlier chapters to identify that the tasks of customizing solutions might benefit from work augmentation through better data from robotic sensors, information process automation, and analysis powered by AI. These are important, but not enough. Leaders must also consider the higher-level implications for the role of marketing work in the broader ecosystem of teams, units, divisions, and so on.

Working from the Inside Out: The Automation of Cancer Treatment

Recall our example of oncology treatment in chapter 3, and the compelling imagery of AI-enhanced reinvention of treatment identification and choice, and social-robotic surgical assistants enabling human surgeons to do less-invasive and more precise surgical procedures. In that chapter, we focused on reinventing jobs to identify optimal human-automation combinations. Table 5-1 (page 135) shows the job-level reinvention of oncology treatment, as described in earlier chapters.

TABLE 5-1

Organizational implications of reinvented oncology treatment jobs

Job-level reinvention

Old job	Automation-optimized job
Oncologist gathers and reviews patient information and estimates the likelihood of cancer.	RPA gathers and integrates patient data in real time. AI reads the data, assesses the likelihood of cancer risk, and provides the oncologist with a preliminary score.
Oncologist orders diagnostic tests, analyzes the results against knowledge of medical research findings, and makes a cancer diagnosis.	Cognitive automation and AI (such as IBM's Watson for Oncology [WFO]) scans millions of pages of medical literature and makes a diagnosis based on constantly upgraded algorithms.
Oncologist evaluates alternative treatments and decides what treatment the patient will receive.	WFO analyzes data more quickly and completely and makes recommendations on the more typical cases. Oncologist makes recommendations on cases that are unfamiliar to WFO.
Oncologist carries out surgical procedures.	Routine surgical procedures carried out by AI and sensor-equipped machines. Oncologist decides what tasks to delegate to the machine. Oncologist carries out the nonroutine tasks.
Oncologist leads the diagnostic and surgical team, serving as a hub for information and decision making.	Team can access data from multiple systems, supported by AI and RPA. AI alerts the team to exceptions requiring attention and tracks the behaviors of other team members.

The Star Model of Organization Design

There are many frameworks, each with its own advantages and disadvantages. To structure our analysis of the organizational implications, we use the star model of organization design. This framework can help describe the organizational implications of work-automation optimization. The "star" model was formulated by Jay Galbraith and refined by thinkers such as Amy Kates, Greg Kessler, Susan Mohrman, Christopher Worley, Edward Lawler, and Stu Winby. Galbraith described the model this way:[3]

- Strategy: The strategy specifically delineates the products or services to be provided, the markets to be served, and the value to be offered to the customer. It also specifies sources of competitive advantage, or capabilities.

- Structure: The structure of the organization determines the placement of power and authority in the organization.

- Processes: Information and decision processes cut across the organization's structure; if structure is thought of as the anatomy of the organization, processes are its physiology or functioning.

- Rewards: The purpose of the reward system is to align the goals of the employee with the goals of the organization. It provides motivation and incentive for the completion of the strategic direction.

- People practices: Human resource policies—in the appropriate combinations—produce the talent required by the strategy and structure of the organization, generating the skills and mindsets necessary to implement the chosen direction.

Galbraith could hardly have imagined the recent advances in work and automation. The star model was developed and typically applied to traditional organizations and work done in traditional jobs through employment relationships. However, these organizational elements can also apply to the emerging world of work and organizations. As we described in our book *Lead the Work*, deconstructed and decoupled work changes the very definition of fundamental ideas such as capabilities, structure, processes, metrics, and HR practices. The framework in this book suggests that "structure" and "process" are made up of increasingly deconstructed and decoupled tasks that are constantly reinvented and optimized to account for automation, as well as alternative work arrangements such as gigs, contracts, projects, tours of duty, and so on. It means that jobs and organizations are constantly evolving and reinvented. Table 5-2 (page 138) describes examples of the necessary reinvention of the organization, using the "Star Model" elements.

Surgeons Are No Longer Godlike

An oncology surgeon meets with hospital administrators to request more discretion for the surgical team, as they use diagnostic information and recommended surgical procedures produced by the robotic surgical assistant. The team likes the reinvented job that gives them deep patient information at their fingertips and using the robot for mundane things like opening and closing incisions. However, they bristle because now their reinvented jobs require them to conduct treatments and surgical procedures that follow the best practices identified by cognitive automation. Sometimes the surgical team believes different procedures would work better. Sometimes the cognitive automation recommends that the robot do a procedure because it has been proven superior, but the team members believe they are more adept. (See table 5-2 for the implications of reinvented oncology treatment jobs.)

Before automation reinvented these jobs, only the surgical team had access to information within the operating theater and personal expertise about patient reactions to surgical techniques. Before automation, surgeons were hired and rewarded for their personal expertise and unique capabilities to choose and perform complex surgery. Administrators

TABLE 5-2

Organizational implications of reinvented oncology treatment jobs

Organization-level reinvention

Organization-al element ("star" model)	Automation-optimized organization
Strategy	The role of a surgical center shifts from "providing the best surgical cancer treatment" to "giving doctors and patients the ability to make the best decisions about cancer diagnosis, prevention, care, treatment, and lifestyle."
Structure	• Surgeons become the "pilots" of automated and AI-driven technology. New jobs combine programming and teaching AI with deep patient treatment. • Power shifts because surgeons who previously held exclusive knowledge, expertise, and authority now share them with developers, programmers, and remote scientific experts. • Hospital administrators, boards, and external regulators now have firsthand access to information and diagnostic results that they previously could receive only through surgeons and operating-room staff. Now, those in nonsurgical jobs are far more expert about the operating room, equalizing their power and authority relative to surgeons and staff.
Processes and lateral capability	• Treatment and surgery decisions are now informed by cognitive automation and fueled by databases, sensors, and collaborative robotics. • Operating-room staff must work closely with equipment and database technologists. • Those whose jobs include providing treatment and those whose jobs include maintaining, analyzing, and evaluating information must collaborate, requiring lateral organizational connections and accountability. • Trust between treatment givers and information and technology managers becomes pivotal. • The process of patient communication and care includes seamless connections with caregivers and analysts, programmers and AI teachers.
Metrics/rewards	• Metrics formerly evaluated the success, cost, and risk of surgical care and recovery. Now, they are reinvented to include the patient experience, the validity of patient and staff decisions, and the use of the most appropriate, advanced, and evidence-based information and options for prevention, treatment, and recovery.

TABLE 5-2 (continued)

Organizational implications of reinvented oncology treatment jobs

	• Surgical team rewards now reflect the success of the entire team, including automation designers and supporters.
	• As software developers and data scientists become more pivotal, their rewards increase, and uniquely "human" tasks command a greater reward premium. Formerly high-paid but routine tasks decrease in value and rewards.
People practices	• Continuous learning and adaptive flexibility become key selection factors for surgeons and the team.
	• Hospital leaders create perpetual learning and work reinvention systems that include not only new skills but also the psychological capability to accept changes in power and accountability.
	• Recruitment and selection emphasize new hybrid capabilities that integrate medical and technological capacity.

relied on the perceptions, expertise, and information that surgical teams shared. Before automation, the only option for administrators was to give the surgical team maximum freedom to operate and make decisions, perhaps bounded only by broad guidelines based on cost or legal liability.

After automation reinvents these jobs, the administrators have access to comprehensive information about surgical practices, patient reactions, and the best practices of thousands of other surgical teams using the same AI and social-robotic equipment. The surgical team is now just one of many pivotal groups. In many cases, information analysts and technology designers and programmers have equally valid and important perspectives. As the surgical team evolves to become pilots of the AI and robotics, it is no longer optimal that they have the exclusive authority to recommend surgical treatment. The reinvented jobs of data analyst, programmer, or technology expert now have an equal voice in such decisions. That was unheard of in the past.

Automation has not just reinvented all of these related jobs; it has created an important challenge for the surgical center's leaders and organization designers. They must balance these organization factors against the benefits and costs of work-level automation. Perhaps leaders should forgo some of the opportunities to reinvent jobs with automation that we described in chapter 4 and in tables 5-1 and 5-2, because the organization-level drawbacks are too serious. Perhaps the pace of automation must be slowed to allow surgical teams time to understand and acclimate to these new roles. Perhaps the optimum moment to implement automation is after the hospital has recruited and hired programmers and AI experts who also have surgical experience and qualifications, because they can bring necessary credibility and trust to discussions with surgical teams that traditionally held all the power and authority.

The New Organization

Automation reinvents organizations, just as it reinvents jobs. Such reinvention includes virtual teams, agile/SCRUM, holacracy, flat and lean organizations with self-managing workers, and the Haier example of the organization as a hub for microenterprises. Automation makes it possible and perhaps necessary to reinvent the organization and supports faster and more radical experiments. Smart software supports flexible staffing and human resource planning. Data analytics monitor performance and predict resource requirements. Smart communication software enables virtual collaboration. Augmented reality simulates in-person interactions. The IoT conveys vital customer and user data directly to the workers, with no supervisor.

But not only organizations are redefined through automation. The meaning of leadership and the role of leaders and followers are also reinvented. That's the focus of chapter 6.

SIX

The New Leadership

Democratic, Social, and Perpetually Upgraded

In *The Inevitable*, author and cofounder of *Wired* magazine Kevin Kelly describes twelve disruptive technological forces.[1] One is "becoming," in which products, services, and relationships are perpetually both obsolete and upgraded. The year 2017 was the tenth anniversary of the iPhone, which by then had established a familiar pattern: as soon as a new one emerges, it's the hottest thing on the market, and the old version is dramatically less valuable. This trend of "becoming" affects work and organizations, too. Of course, organizations and leaders must be agile. An Accenture report recounted that CEOs listed "becoming agile" as their third-highest business priority, noting that "HR will enable a new type of organization—one designed around highly nimble and responsive talent."[2] Leaders, workers, and HR systems must prepare for this new world of perpetually upgraded work, just as they learned to deal with perpetually upgraded iPhones and other technology.

Leaders, workers, and policy makers understand this, but often only in very generic ways. A recent Genpact global survey of five thousand people revealed that only 10 percent strongly agreed that AI threatens their jobs today, but 90 percent believed that younger generations would need new skills for success.[3] It's easy to be lulled into a false sense of security, thinking that automation will affect only future generations or workers in jobs other than yours. Leadership will require encouraging a constant reexamination of work and jobs that recognizes that automation will reinvent jobs and then precisely identifying the implications and the collaboration needed to optimize it. This chapter provides guidance about that new leadership, optimized for the future world of agile, automated-enabled, and reinvented jobs.

Our framework helps you optimize work automation and diagnose and anticipate the reinvention of jobs. That means reinventing the role of human workers and the organization. It means that the role of leaders will also be perpetually reinvented. The distinctions between the leader and the follower are increasingly blurry, in part because automation makes information that was once reserved only for leaders instantly available to workers, customers, and constituents. Leadership is no longer exclusively for those in certain jobs that formally include formulating strategy, having vision, fostering communication, engaging followers, and providing a role model. Organizations like Haier and Zappos aspire to organization designs that have few or no managers, meaning leadership is whatever empowers those closest to the customer, including automation. Leadership development and succession must increasingly reflect an uncertain future, because constant reinvention makes it impossible even to describe the future organization, strategy, and value propositions that will be led.[4]

So, future leaders must optimize a changing mix of deconstructed tasks that are constantly being reconfigured to exploit new human-worker arrangements (gigs, contracts, projects, employment, and crowdsourcing) and new combinations of human and automated workers. Leaders and workers must freely exchange information, even when that information means that some tasks will be removed from human workers, and their jobs will go away or change. Leaders must lead the work, not only the employees, creating an ecosystem of current and potential workers who are willing and eager to engage, as they adjust to perpetual change.

As a leader, you must become adept at deconstructing and reinventing work and your organization. You must also share frameworks to optimize work and automation, because often your followers will be the first to realize new job-reinvention opportunities.

Leading Perpetually Reinvented Work

A recent Willis Towers Watson Future of Work study found that global companies expected work automation to increase from 7 percent in 2014 to 22 percent by 2020. The same survey found that companies expected work done by "nonemployee talent" to increase from 16 percent to 23 percent between 2017 and 2020. If you think of reinventing work using both new human work arrangements and automation, it suggests eight options:

- Traditional employment

- Outsourcing

- Free agents

- Alliances

- Talent platforms

- Volunteers

- Robotics

- Artificial intelligence

In previous chapters, we showed that when reinventing jobs using automation, the optimal patterns are not revealed by seeking binary solutions, such as "contractors versus employees" or "robots versus humans." Instead, jobs must be deconstructed and the tasks done with the best option, and then those optimized work elements are reconstructed into reinvented jobs that may include contracts, gigs, and automation solutions. Leadership involves engaging workers as collaborators, helping to perpetually track how their work is evolving, and being willing and confident to identify new alternative approaches.

Even traditionally human-centric professionals like lawyers and accountants are finding their jobs reinvented as RPA and cognitive automation take on repetitive, cognitive tasks, and the best human workers increasingly find platforms and freelancing a more preferable arrangement to traditional employment. First, the work of junior professionals is substituted by RPA that can more quickly and precisely find relevant legislation or technical information. Then, the work is reinvented, as AI can do much of the analysis itself, and the job for humans shifts from analyzing financial statements to teaching AI to understand financial statements. Leadership means balancing the risks and returns of full-time employment versus shorter-term and less-traditional engagements and automation. Overemphasizing traditional employment risks creating jobs that soon require painful and disruptive adjustments. Overemphasizing temporary engagements or too aggressively replacing human work with automation risks creating a workforce that is insecure, resentful, disengaged, or unavailable. Overemphasizing automation risks embedding biases and limitations into the technology that make it slow to adjust to unique new problems.

The five transformative changes that redefine leadership are:

- Mindset: From "learn, do, retire" to "learn, do, learn, do, rest, learn . . . repeat"

- Ability: From employment qualifications to work readiness

- Reward: From salaries for permanent jobs to flexible total rewards for deconstructed tasks and work arrangements

- Deployment: From job architecture and movement between jobs to work architecture continuously matching capabilities to tasks

- Development: From career ladders based on fixed jobs to reskilling pathways based on tasks and reinvented jobs

We will discuss each transformative change in depth.

Mindset

From "learn, do, retire" to "learn, do, learn, do, rest, learn . . . repeat": In his 1970 bestseller *Future Shock*, Alvin Toffler said, "The illiterate of the 21st century will not be those who cannot read and write, but those who cannot learn, unlearn, and relearn."[5] His observation becomes more relevant every day.

The great twentieth-century giants like General Motors and Ford grew by aggregating individual craftsmen in cottage industries into jobs in centralized factories. For decades, the careers that created income and development during employment and retirement comprised those jobs. Predictability and stability allowed schools to train talent before employment, organizations to add additional long-term skills, and careers to progress consistently through job families organized into functions like R&D, manufacturing, HR, finance, and sales. Authority and accountability predictably progressed from individual contributor to supervisor to executive. This linear progression worked with predictable and relatively stable economic growth, and supported organizations of global scale and hundreds of thousands of employees. Perhaps the most vivid icon of this stability was defined retirement pension benefits and medical plans, made possible by predictable growth and shorter life spans.

Of course, modern reality could not be more different. The volatile, uncertain, complex, and ambiguous modern world is amplified by the technology convergence and global transparency of the fourth industrial revolution (see chapter 3). This reality is often recognized in decisions about money, technology, innovation, customers, and markets. We see it in the shrinking half-life of skills and the relentless reinvention of jobs. These changes in work combine with societal trends such as increased life expectancy, ubiquitous virtual connectivity, social media, cyber threats, and income inequality.[6] Table 6-1 summarizes some of these changes.

Past generations could rely on rewards from a pattern of "learn, do, retire." That no longer holds. Jobs and professions have a shorter half-life, even as the duration of our work lives increases. The World Economic Forum estimates that 65 percent of children entering primary school will ultimately end up working in jobs that don't yet exist and in professions that are very different and constantly changing.[7] The new pattern will reflect a series of careers, built upon projects and shorter tours of duty in each organization.[8] That requires a mindset more like "learn, do, learn, do, rest, learn . . . repeat."

In chapter 5, we described how these work changes are enabling and requiring new organizational forms, supported by constantly reinvented jobs. However, the same work and organizational evolution also has significant implications for leaders. Leaders must adopt and encourage

TABLE 6-1

How work is evolving

From	To
Stable, predictable jobs; career and reward progression	Perpetual reinvention of jobs; unpredictable and changing careers and variable rewards
Economic and social predictability	Volatility, uncertainty, complexity, and ambiguity in economic and social environments, amplified by technology convergence.
Life expectancy of 65 years	Life expectancy of 85 years
Stable professions and work patterns despite changing technology	Shrinking half-life of professions, skills, and work patterns

JD's Perpetually Upgraded Mindset About Development and Careers

JD, China's largest retailer, is the first Chinese internet company in the *Fortune* Global 500. As of 2017, JD's compound annual growth rate was over 150 percent during the previous thirteen years, which made it the fastest-growing leading e-commerce company in the world. One hallmark of JD's value proposition is its strict zero-tolerance policy for counterfeit goods. It has the largest e-commerce logistics infrastructure in China, covering 99 percent of the country's population, with over 92 percent of orders delivered on the same or next day.

Obviously, an advanced supply chain, logistics, sourcing, and customer service are pivotal processes to ensure JD's unique value proposition. JD is using AI-enhanced intelligent logistics, supply chain, and customer service to deal with increasing complexity and growth, and to improve operational efficiency.

At the operational level, JD launched the world's first fully automated warehouse in 2017, where goods can be loaded, stored, packed, and sorted automatically and intelligently. In customer service, its self-developed intelligent robot, JD JIMI (instant messaging intelligence) responded to nearly 90 percent of customers' first-time online inquiries. JD also developed intelligent supply systems to choose commodities, optimize prices, and monitor inventory based on historical data analysis.

In this situation, demand for some traditional jobs decreases, but demand for work combining humans, automation, and AI increases. Done properly, new technology frees the workforce from dangerous or repetitive work, and enables workers to focus on core responsibilities or develop new capabilities. However, that requires an approach to the work that recognizes the reality of perpetual upgrading, and engages leaders and workers in a constant dialogue, guided by data. It means that JD leaders must forecast future workforce capabilities and help employees make the transition.

JD chose to tackle these changes by initiating its Project Z, which created a cross-functional virtual team led by the JD HRI (Human Resources Research Institute). Project Z's goals are to study, monitor, and forecast how new technology will influence organization and talent, recognizing jobs that might disappear or be transformed to fit changing technology, trends, and JD strategy.

Based on its research, JD HRI predicted that the number of JD employees will actually continue to increase, and no reduction would be necessary in the next three years. That's partly because the rapid business growth will offset any impact of automation on workforce reduction. The research also revealed that workforce changes will happen faster in some areas than in others. To achieve these insights, JD HRI developed its own technology-transition workforce forecasting model, called the "JD Workforce Weather Forecast." This forecast allowed JD to prioritize the focus of its study and its transition approaches on the most pivotal work transitions.

The matrix shows the four categories of work-automation transitions, according to the speed of their arrival and their impact on the work. The jobs in the "storm" and "plum rain" categories get top priority for rapid and larger workforce transition investments.

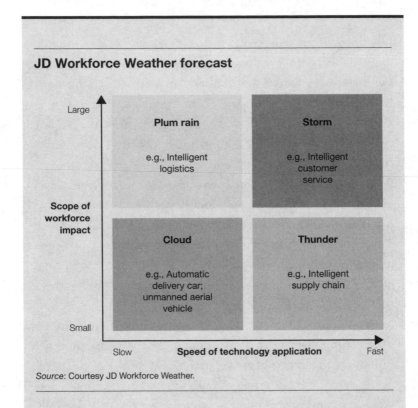

Source: Courtesy JD Workforce Weather.

For technology-transition under each category, JD identified a road map of how the workforce will change in three
years and formulated a talent development planning system
that relies on open collaboration between business leaders
and workers using perpetually upgraded talent forecasting
and development:

- In intelligent customer service (the "storm" category),
  there is rapid technology application and a large impact
  on the workforce. Here, JD plans to further increase
  the application of intelligent robots in responding to
  customer inquiries. Since 2017, JD has been developing

and implementing a comprehensive skill enhancement plan to address the upcoming workforce replacement, including reclassifying work responsibilities and upgrading employee skills to a higher level, transitioning employees to newly created jobs, such as an "AI trainer" who helps robots with data markup, data mining, knowledge integration, and so on.

- In intelligent logistics (the "plum rain" category), there is a large impact on the workforce, but relatively slower expansion of work automation due to limitations such as equipment, production, automation maturity, and so on. In this category, JD initiated its cloud ladder plan, training its warehouse workers with greater skills to support new professionals in central control, automatic warehouse systems, and equipment operation and maintenance. In addition, JD launched unmanned aerial vehicle pilot training.

- In intelligent supply chain (the "thunder" category), there is rapid automation, but only a small impact on the workforce. Here, JD is designing new job levels and career development paths that look beyond the three-year horizon and are supported by training for procurement professionals. The training was designed by predicting how the work will change after the systems go live, beyond the three-year horizon. In the intervening three years, JD HR will work with business units to forecast staffing for procurement and help its workforce adjust to the forecasted new work reality.

a mindset of constant retooling, supported by necessary changes in the very ideas of work, culture, and values. Leaders must pursue their careers this way and require, support, and encourage workers to adopt a mindset of continuous development. Instead of saying to workers, "This is what you will become," leaders must say, "You start with these skills and capabilities, and our work is to collaborate to refine and enhance them." For example, leaders can start by deconstructing competencies, skills, and capabilities, and distinguishing those with short-term value from enabling skills with longer-term value that underpin the ability to thrive in constant change. Enabling skills include things like critical thinking and global mindset. They are very different from technical skills that have received such significant social and policy attention, such as mathematics, computer coding, and so on. Leaders once expected workers to acquire technical skills before employing them. Increasingly, leaders select workers less on technical skills and more on enabling skills. Leaders must identify and nurture the ability to learn and apply a changing array of technical skills. Acquiring and changing your technical skills will be a routine, repetitive exercise that occurs as work and jobs are constantly reinvented and perpetually upgraded.

The Global Talent 2021 study by Oxford Economics, in collaboration with Willis Towers Watson and a number of other corporate and academic partners, showed that "agile thinking and relationship skills" were rated by leading executives as two of the most important future skills.[9] This includes the ability to consider and prepare for multiple scenarios, manage paradoxes and balance opposing views, innovate, and co-create, and brainstorm. They were rated as far more important than specific technical skills.

This reinforces our descriptions of how work automation reinvents jobs. Manufacturing workers will have technical skills in machine tool repair and operating a precision drill. They will also have the skills of collaborating with colleagues, diagnosing problems, and understanding how work fits with broader processes and contributes to the total solution. As work is reinvented, automation may take on 60 percent of the tasks, such as operating the precision drill and removing and installing tools. Does this mean human workers will be replaced? Probably not, but their value is now in training the automation and collaborating with the automation to diagnose problems and invent solutions. At first, the human workers' prior technical skills will support the tasks of training the automation. As reinvention evolves, the automation will have learned enough to operate on its own. The workers' enabling skills in critical thinking and analytical ability will help them add value by analyzing and repairing the robot.

For leaders, this means creating a relationship and a work system where the human workers feel safe and motivated to report when they envision new automation applications that might replace some of their tasks. Leaders must show that, as automation progresses and jobs are reinvented, they can be trusted to help the human workers adapt or move to other organizations in a humane and exciting way.

Ability

From employment qualifications to work readiness. Employment qualifications are often defined using technical skills. Employers and policy makers lament the fact that companies can't find workers with the specific skills for today's work. Governments, employers, and educational institutions try to identify and provide those skills to reduce the gaps. However, as value shifts from technical to enabling skills, education, training and learning must retool accordingly.

Workers can increasingly acquire technical skills quickly and cheaply, outside of employment or educational institutions. Lynda.com, the world's largest online learning resource, contains all the courses needed to qualify a C++ developer, who can acquire those skills in as little as fourteen hours of focused learning. Lynda offers learning pathways for many technical skills (Python, Java, iOS10, graphic design, 3D animation, network/infrastructure administration, etc.); Lynda is only one such source. Online learning sources also give workers a clear understanding of adjacencies. For example, graphic design and 3D animation are adjacent skills, because the work of graphic design can be upskilled to 3D animation work. On the other hand, although they share some common attributes, Python programming skill is not closely related to graphic design, because it takes many more hours and a longer pathway to go from Python to 3D animation, even though 3D animation may be done on software that uses Python programming.

Leaders must understand such relationships and offer workers the most effective learning path, even if that means leaving their organization. Learning is integrated with rewards by combining an online marketplace or talent platform like Upwork with the online learning solutions of Lynda. Now, a Java programmer on Upwork can earn $40 per hour, but can easily see that Android programmers earn $90 per hour. Lynda.com shows them that with four additional courses over fourteen hours, they can become certified Android programmers. When the price and alignment of such skill pathways is so apparent, leaders can and must shift their focus from searching for candidates who are fully prepared to search for those who are optimally close to being qualified and strategies to optimize the work and worker to achieve the most efficient match.

Automation plays into this in two ways: it accelerates the continual evolution toward new skills, and it provides the means to acquire and demonstrate those skills. As technical skills become increasingly easy to acquire and change, leaders will expect educational institutions and their organization's long-term training and development to reinvent with a focus on enabling skills. Typically, such skills are codified only in proprietary corporate skill inventories or competency systems. So, they are not transparent and transportable. They seldom appear on a university diploma or certification, for example. Future organizations may need to reinvent that by supporting platforms that track and report enabling skills, as technical skills are now.

Leaders must enable relationships with workers that span multiple engagements of different kinds, punctuated by workers leaving to acquire new skills through education or work with other organizations. Leaders must learn to identify and track enabling capabilities, even among workers who may not currently possess the formal certifications or degrees in required technical areas. The best talent may well be a worker with an amazing ability to learn and see connections, but whose technical qualifications don't yet match current needs.

Reward

From salaries for permanent jobs to flexible total rewards for deconstructed tasks and work arrangements. Typically, organizations value and reward work by combining tasks into a job and then surveying a market of rewards for comparable jobs in comparable organizations. Data provided by online sites such as salary.com and LinkedIn has increasingly modernized the process. Certainly, a worker's rewards reflect personal attributes like experience and "hot skills" like Python programming, but typically the rewards are attached to a job. This is even true when workers are engaged through arrangements other than regular full-time employment, because the requisitions for contractors or freelancers are often based on jobs.

Yet, step one of our framework requires deconstructing jobs into work tasks and then constantly reconfiguring and reinventing new jobs. Step two estimates the ROIP of tasks. So, in the future world of perpetually reinvented jobs, future rewards must reflect the ROIP of tasks. Then, they must reflect the cost, risk, and productivity of human workers versus automation for those tasks, and the optimal combination of humans and automation.

How can work be valued in this new world? Traditional job-based market surveys are incomplete and inefficient because needed information is proprietary, resides in different places, and is opaque or difficult to get. So, the reward market that relies on pricing jobs moves slowly such that the pricing base is stable. A market for deconstructed tasks can be much more efficient, as talent platforms already show. If you want to engage an Android app developer, you can go to Upwork, Appirio, or other such platforms, and immediately find an array of developers and current pay rates, with the range reflecting their different past performance ratings, experience, and current expertise. The prices for the tasks of Android app development, and even the language to describe the work, all change quickly, as skills change, as parts of the work are automated, and as workers respond by changing their capabilities. Because the unit of analysis is the deconstructed work task and not the job, these markets support a higher volume and velocity of transactions. As workers and employers discover work tasks that they can substitute or augment with automation, or as automation creates entirely new versions of the work, a market freed from job descriptions can adapt more easily and quickly. This adaptation reflects more than merely how much a task is valued. The market can also differentiate rewards to reflect location, independence, continuity, reputation, and even how the work supports missions such as environmental sustainability, social justice, and so on.

How does this reward system change leadership? Leaders and workers will more constantly negotiate and renegotiate the nature and rewards of work, because work and jobs will be perpetually reinvented. Those reward negotiations will focus on tasks that are deconstructed and reconstructed to optimize automation as well as different human-worker arrangements. Both leaders and workers will increasingly have access to the same information, a far different situation than that in which only the leader knows organizational pay levels and only the workers know their true capabilities and alternatives.

The example of the reinvented job of an oil driller in chapter 1 illustrates the evolving rewards. Recall how the job is changing, shown in table 6-2.

TABLE 6-2

Changes in the job of oil driller

From	To
Analog gauges and operator expertise	Digital, interactive cockpits with automated functions
Primarily physical work	Primarily mental work augmented with automation
Focus on rig-centric control	Shared control with centralized operations center
High labor intensity, low skill premiums	Lower labor intensity, higher skill premiums
Significant variation in operating performance and predictability of maintenance	Greater predictability of maintenance events and much lower performance variation through sensors, AI, and analytics

Traditionally, leaders would value this work by comparing their organization to a market of other companies that employ workers in the job of "driller," described in the left column of the table. The market data has little relevance to the reinvented job shown in the right column. How should a leader value the new job? Earlier chapters showed that the work optimization choices will increasingly be based on distinctive combinations of deconstruction, automation compatibility, ROIP, automation availability, and organization considerations. Because every organization is reinventing its jobs differently and rapidly, there will no longer be a sample of comparable jobs in other organizations. So, no one salary survey provider may be able to give market data for the new job. Of course, eventually, this reinvention process may produce some common and uniform work-automation combinations across organizations, but that will likely take too long. Leaders will need to deconstruct and reconstruct the work far faster than traditional salary surveys can reflect in their jobs.

An alternative way to establish the market value for the new job might be to estimate the proportion of the work for each task:

- Cockpit monitoring and response (25 percent of your time)

- Coordination with operations center (30 percent of your time)

- AI training and system innovations and upgrading (20 percent of your time)

- Change leadership and stakeholder engagement (25 percent of your time)

Then, you estimate a market value task by task. You might consult a talent platform that prices the component tasks of the new job. You might scan databases of community colleges for the salaries of certified graduates. You establish prices for the tasks, not the job, so you are not limited to organizations in your industry. You can tap data from a wide variety of industries, including aviation, mining, and transportation. You might end up with the following average prices on a per-hour basis:

- Cockpit monitoring: $45

- Coordination with operations centers: $30

- AI training and system innovation and upgrading: $60

- Change leadership and stakeholder engagement (based on freelancers who do change management)

As you do this, you realize that you could execute the human work tasks with combinations of freelancers, con-tractors, and regular employees. However, if you decide to reinvent these tasks into one job, and assume a two-thousand-hour work year, the estimated market value for your new job is $114,500.

How do you consider the enabling skills of your current employees in the driller job, such as learning agility, critical thinking, and global mindset, as well as organizational history and knowledge that is critical connective tissue to ensure the various activities are appropriately integrated? These skills are intangible because there is little market data. So, you might attach a 20 percent premium for these intangibles, bringing the market value of the new job to $137,000.

As talent marketplaces like Upwork grow in scope and scale, the quality of data will only improve over time, making such analysis increasingly more common and straightforward.

Deployment

From job architecture and movement between jobs to work architecture continuously matching capabilities to tasks. Traditionally, talent is deployed to jobs, using job architectures. A job like "software development engineer" exists in a job family like software development that contains multiple engineering jobs and levels. Each higher level has greater scope, impact, and responsibility. Several job families like software development and network design are grouped into job families like engineering. Organizations hire and develop workers through jobs and job families.

This can be costly and inefficient. Moreover, as we have seen, perpetually upgraded or agile work is reinvented too quickly for such fixed and job-based architectures to keep up.

Deploying workers from one job to another is just too blunt-edged to capture the nuances of deconstruction, reinvention, and work automation. When the unit of analysis is the work tasks, more agile knowledge architecture can use data from multiple sources like LinkedIn and Upwork to match worker capabilities to work.

Recall the example of the oil driller. It may be possible to map the adjacency of those who did the old driller job to show them precisely what skills they need for the new job and how to get those skills. Some of the previous drillers can be deployed to the new job if they acquire the needed skills. Or, some of the former drillers may work part-time on some tasks in the new job, while also pursuing other work within the organization. Knowledge architecture, often powered by AI, can mine information on talent inside and outside the organization and match it to demands as they emerge. Workers can be borrowed for a project or set of tasks requiring skills and domain expertise for the duration the work demands.

As a leader, your relationship with the work and workers changes fundamentally. You oversee deployment options that cross the organization boundary, and your role is not so much to match workers in one job with their next job, but rather to optimize worker development to fit perpetual work evolution. You will more frequently look for workers who are close but not necessarily perfect matches. You will deploy workers to projects so they can develop targeted skills. You will use the language of jobs less often and the language of deconstructed tasks, automation, and workers' capabilities and desires more often.

As job architectures enabled the organizations of the second industrial revolution, the deconstructed task and knowledge architecture form the basis for the more networked ecosystems of work that will likely characterize the fourth industrial revolution.

Development

From career ladders based on fixed jobs to reskilling pathways based on tasks and reinvented jobs. How do we then ensure the continued reskilling of humans? In the past, stable economic environments and technology allowed development to happen within specific professional domains and a single organization. Accountants built a base of technical skills and then took additional classes and a progression of jobs to expand them. They might have started a career doing financial reporting in a corporation and then shifted to working as an auditor with a large accounting firm, leveraging knowledge of US accounting laws and regulations. Then, they might have shifted into consulting or management accounting, building on their knowledge to add expertise in accounting principles outside the United States. This journey followed a predictable and stable path to build and enhance technical skills that were easily verified and tracked.

In the future, such technical skills will change more quickly, often being replaced or altered as the work combines with automation. Moreover, those changes will be difficult to predict far in advance. Enabling skills will last longer, but workers will need to develop technical and enabling skills and to adjust to career paths and learning pathways that change quickly.

As a leader, you will play a key role in whether the future of work and automation means the demise of learning and development within organizations, or the birth of a more precise, comprehensive, and boundless approach. This new approach will better account for the whole worker, rather than only those attributes that matter for a particular job or that are included in your proprietary organizational competency model. It's possible to imagine future leaders as unconcerned with worker development, relying on the workers themselves to navigate a more transparent system of platforms and online career communities like LinkedIn. However, leaders have the opportunity to create unique worker engagement by guiding them through a connected and evolving array of development options, informed by more open work architecture. Leaders in future best places to work will likely become just this sort of skilled guide.

Work-automation optimization through deconstruction and job reinvention will provide leaders with greater insight into where and how automation will replace human labor for specific tasks. Also, reskilling will increasingly rely on enabling skills, not technical skills. The new reskilling pathways will map enabling skills through many different types of work that will span different professional domains, work arrangements, and organizations.

Accountants, in addition to their technical accounting skills, might also possess enabling skills: a global mindset, a strong process and method orientation, caution and risk aversion, and a learning orientation.

Reskilling pathways would track how these skills support the full development path and the new technical skills acquired along the way. Now, an accountant's career might include managing an oil rig in Saudi Arabia, leading the actuarial function at a global insurer, or operating as an independent quality assessor for a major pharmaceutical company. These jobs seem very far from the typical career path. What do these different types of work have in common? They each require the following enabling skills for success:

- Oil rig manager. A global mindset is needed to manage a team of workers from around the world. The process and method orientation ensures the integrity of highly repetitive, process-based work. The enabling skill of caution leads to success when a small mistake can have devastating consequences.

- Actuarial leader. The global mindset is evident in leading a global function. The process and method orientation supports maintaining the integrity of determining reserves, evaluating claims, and so on. Caution and risk aversion enhance performance on risk diligence.

- Independent quality assessor. A global mindset is not used to supervise a global team, but rather for evaluating processes and products in many different countries. The process orientation now supports creating consistent, repeatable processes that can be audited and verified. The enabling skills of caution and risk aversion are in play for establishing appropriate risk tolerances for deviations from established standards.

Skills of the Successful Leader of the Future

As jobs and organizations perpetually reinvent themselves to optimize work automation, leaders will evolve from hiring talent and delegating tasks to orchestrating evolving work delivered by automation and many different human-worker relationships. Skills such as continuously deconstructing and reinventing jobs and an ability to not only find and nurture technical competencies but enable skills will incur a premium.

Optimizing work automation by perpetually reinventing jobs requires fundamental changes in leadership and leaders' relationships with workers. One of the most important changes will be in the transparency with which leaders and workers address constantly reinvented work. The most agile organizations must have everyone—workers and leaders alike—willing and able to candidly share what they know about how work is changing and to reinvent it. That will take courage on the part of leaders.

When John Boudreau interviewed former Secretary of Commerce Carlos Gutierrez, the secretary observed that a competitive and agile US economy depends on a competitive and agile workforce that can identify evolving work opportunities and the evolving pathways to prepare for them. He recalled one of his toughest decisions while CEO of the Kellogg Company—to close the Battle Creek manufacturing plant in 1999.[10] The original plant was an icon within the company, but its processes were obsolete in an age of modern manufacturing. Gutierrez and the Kellogg team did what they could to treat workers humanely, but there were limits to the pathways they could offer, particularly for workers who were unable or unwilling to relocate. They informed the workers about the closing shortly after the decision had been made.

Boudreau asked Gutierrez how much advance notice he had had before the plant closed. He said that several previous CEOs had seen the inevitability of the closing, but the daunting prospects of disrupting workers and the community had delayed the decision. Gutierrez felt it was his duty not to pass it along to the next CEO.

Currently, workers faced with a plant closing may have more options, including living in Battle Creek but using virtual tools and freelance platforms to find future opportunities. In the increasingly data-rich and agile work world, workers and leaders should perpetually prepare for inevitable work obsolescence. Might future events like a plant closing be less shocking in the new world of agile pathways?

That requires a new mindset. Gutierrez suggested that even today's leaders, when faced with work obsolescence and disruption, will instinctively wait to engage workers after a tough decision is made. He said, "Looking back, my team and I had a choice about how early and transparently we would share our knowledge that the plant and its work would soon change drastically."[11]

Leaders assume that if they reveal disruption too early, it will produce worker stress, contentious labor union or community reactions, and departures of key employees. Why risk starting an unpleasant conversation earlier than necessary? Such traditional assumptions must be questioned if leaders and workers embrace agile work and learning.

Every day, leaders and workers have choices about how transparently they share knowledge of how work is changing. Candid and honest conversations about the perpetual upgrading of work provide workers and leaders time and opportunity to adjust, even if it's painful. Are your leaders driven by old assumptions to keep quiet until disruptive change occurs? Or, is HR equipping leaders and workers to transparently perceive, discuss, and prepare for inevitable work changes?

As a leader, you must prepare for all of the changes automation brings. You must also prepare for the ways in which your own job might be automated. We turn to that topic in our final chapter, showing how to apply our four-step process to your own job.

- - - - - - - -

SEVEN

Deconstruct and Reconfigure Your Work

Using the Work-Automation Framework to Navigate Your Personal Work Evolution

Now you understand how the accelerating reinvention of jobs optimizes work automation. Our framework and examples show that it's easy to get caught up in predictions about the end of human work, robots replacing humans, and sensors on every object that create big data. You have seen how cognitive automation can predict human behavior better and faster than any person and teach itself how to beat both human and computerized game players. A quick web search will easily unearth predictions such as a recent *Fast Company* article listing "the 10 jobs that will be replaced by robots," including bank tellers, journalists, and movie stars.[1] Or, a compelling graphic in *Politico* showing a future in which humans care for aging humans, while robots do everything else to move goods and information.

Our framework shows that automation will not replace people anytime soon. More realistically, the future will present constantly evolving combinations of human and automated workers. You can only see the patterns in that evolution through deconstruction, ROIP, and optimized work automation.

How should you prepare yourself? In this final chapter, we explain how to apply our "reinventing jobs" framework to your own work and career. The framework can help you identify more optimal goals for and approaches to your career and work.

The Perpetual Conversation about Agile Work and Reinvented Jobs

We've recommended, in chapter 6, that workers and leaders must have perpetual conversations about how their work is evolving so they can identify ways to upgrade work, reinvent jobs, optimize work automation, and balance human and automated work sources. The only way to keep up with the evolution is to have those conversations constantly. In that way, you are more likely to anticipate the evolution that affects you most, well in advance of a disruptive moment when automation replaces large parts of your job. Your conversations must include not only you, your leaders, and your fellow workers, but also your educational institutions and governments.

The reinventing jobs framework lets you have intelligent, nuanced, and productive conversations. You can get beyond the typical hyperbole about robots stealing human jobs and all workers becoming gig workers. You can now frame the conversation to focus on deconstructing jobs, describing ROIP, and reinventing jobs to optimize the combination of automation and human work.

Where can you start these conversations?

Sometimes your organization will provide the answer. In our previous book, *Lead the Work*, we described how IBM tackled these new work challenges by building a talent marketplace that deconstructs jobs into tasks and presents its workers with the tasks on a platform where they can volunteer and track their accomplishments. IBM's approach entails deconstructing jobs and allowing workers to choose short-term projects. It also includes reinventing jobs by deciding what tasks to keep within a job. Reinvention encompasses permeable boundaries, with short-term assignments that sometimes exchange IBM employees with the employees of clients or partners. This internal talent marketplace not only encourages conversations about reinventing jobs, but requires them. IBM workers and managers must constantly work together to optimize the balance of tasks, regular jobs, and automation.

Sometimes the catalysts for the new conversations will emerge from educational institutions. Eloy Ortiz Oakley, chancellor of California Community Colleges, has noted experiments with alternatives to multiyear degrees. For example, portable and "stackable credentials" allow workers to cycle between learning and earning.[2] They can take short courses to acquire credentials with immediate labor market value, return to work, and then come back to college for additional credentials. All the credentials along the way count toward certificates or degrees, and different combinations of credentials can be stacked to qualify for jobs or compile a certificate.

Sometimes, the catalyst for these conversations comes from nations and governments. For example, every nation is wrestling with the question of how to provide income continuity while workers are acquiring new skills or taking a break. One popular and controversial solution is universal basic income (UBI). A UBI in its purest form is a periodic cash payment to individuals that carries no means test or work requirement. It's touted as a way to allow people to afford basic food and shelter, without requiring an organization to pay them. Workers would be free to pursue the work they enjoy or get qualified to jump to an entirely new profession. Finland launched a two-year experiment with UBI, selecting two thousand unemployed people to receive 560 euros per month (about one-fifth of average income).[3]

A variant on this theme is a hybrid of unemployment benefits and insurance, when workers receive income after automation displaces them to support their efforts to reinvent themselves. The Singapore government provides subsidies and a government-approved list of skills, encouraging workers to acquire the skills and companies to give their workers the necessary time off.[4]

Bill Gates has proposed "a tax on robots"—an employer who replaces a worker with a machine would pay taxes equal to those that the worker would have paid. The idea is to slow the pace of automation and create funding to give displaced workers the tools and opportunities to qualify for other work. The city of San Francisco was actively considering the idea in 2017.[5]

Organizations, educational institutions, and governments are important to a future work ecosystem that effectively adapts to the inevitable reinvention of jobs. Yet, they all rest on a foundation of workers and leaders like you. Every worker and leader has a role in the ongoing conversation, because they are on the front line. They will be the first to see the risks and opportunities of job reinvention. The faster and more alertly they detect them, and the more frequently and transparently they discuss them, the more likely that organizations and society will anticipate work upgrades early enough, and with enough sophistication, to avoid needless and wasteful disruption.

The reinventing jobs framework can guide those conversations.

Reinventing Jobs: A Personal Career Tool

The four-step process and framework can help you anticipate how your own work will evolve and prepare for it. Deconstruct your current and future jobs into work tasks. Then ask how the ROIP and automation are evolving. Where will automation substitute or augment your work? What can you do to reinvent yourself to better fit the tasks that will remain human?

Follow the four steps as they apply to your current and future work:

- Step one: Deconstruct your job. Key insights are revealed in the tasks within jobs.

- Step two: Assess ROIP. The payoff is the return on improved performance.

- Step three: Identify automation options. Choose from process automation, AI, and robotics.

- Step four: Optimize work. Find the right combination of deconstruction, ROIP, and automation.

Then, in the spirit of chapter 5, add:

- Step five: Navigate the organization. Find your place in a digital, agile, boundaryless work ecosystem.

Step One: Deconstruct Your Job

Deconstruct your job into its tasks and consider how each will evolve. To find your work tasks, start with your job description and its tasks, outcomes, and competencies. Also, consider how you actually do the work, particularly if that's different from the formal job description. Describe the additional tasks and add them to your list.

Evaluate each work task against the three characteristics of work elements, to see which may evolve away from traditional employment or toward automation. Are the tasks:

- Repetitive or variable?

- Independent or interactive?

- Physical or mental?

The more repetitive, independent, and physical the task, the more likely it can already be automated or will be soon. Even mental work, if it's independent and repetitive, can be automated by RPA and AI. Physical work will probably be automated by robotics.

Now, write the job description as it might appear in two, five, and ten years. Remove the tasks where automation will substitute for humans. Keep those tasks where automation will augment the work and consider how your work will change with augmentation. Finally, what tasks are humans likely to do for a long time? Which of those tasks are regular employees likely to do, and which might be done through arrangements other than employment?

Once you deconstruct and reinvent your job, you will probably find that your reinvented future job contains a fraction of the tasks that you do today. What work elements might be added to your job as it evolves (i.e., new work that will be created)? Can you reconstruct a job from the work that remains plus the new work? How does that align with your unique skills? Could you do some work through freelance platforms, gigs, contracts, or other engagement models either in your organization or elsewhere?

The Economist describes how future work will combine human workers, automation, and freelance work platforms:

> For autonomous cars to recognise road signs and
> pedestrians, algorithms must be trained by feeding
> them lots of video showing both. That footage needs
> to be manually "tagged", meaning that road signs and
> pedestrians have to be marked as such. This label-
> ling already keeps thousands [of human workers]
> busy. Once an algorithm is put to work, humans must
> check whether it does a good job and give feedback
> to improve it. A service offered by CrowdFlower, a
> micro-task startup, is an example of what is called
> "human in the loop." Digital workers classify e-mail
> queries from consumers, for instance, by content, sen-
> timent and other criteria. These data are fed through
> an algorithm, which can handle most of the queries.
> But questions with no simple answer are again routed
> through humans.[6]

Step Two: Assess ROIP

For each current and reinvented task, how does improved performance create value? The four ROIP curves we described in chapter 2 can help you map the payoff of the work. Figure 7-1 shows the four curves; the vertical axis is the value of performance, and the horizontal axis is the level of performance.

You can often find clues to the ROIP of work tasks in discussions with your organization about your performance, key performance indicators, and so on. Conversations about performance, goals, rewards, and development often rely on a shared understanding about the tasks that make up the job, and how those different tasks pay off.

With ROIP identified, consider the implications for work and automation. For the tasks that add ROIP by avoiding mistakes, could robots, RPA, or cognitive automation better

FIGURE 7-1

ROIP for a full range of potential work performance

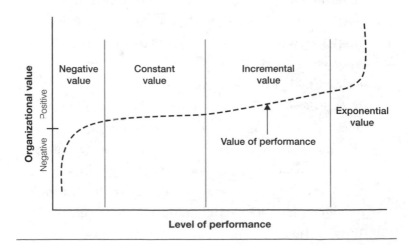

reduce human error? If the tasks with error risks are automated, might you be free to spend more time on the tasks that produce incremental value?

For the tasks where different performance approaches add very little value beyond a certain minimum standard, is it possible that AI could learn the best way to do the work? Or will AI help train or nudge workers toward a common way to perform? If you could expect this work to be done consistently, would that allow you to stop wasting time on unproductive variations?

For tasks that add incremental value, will automation help you move up and to the right on the ROIP curve? For example, if you add incremental value when you interact with customers, will automation improve your interactions by giving you better information about the customer? Will AI essentially make you and other workers equivalent to today's top performers? Or, is your greatest value in training the AI on the most successful customer-interaction patterns?

Finally, for tasks that create exponential ROIP, how will automation create whole new kinds of work? For example, if your value is in understanding the subtleties of how equipment operates or how parts of a process integrate, then when automation takes on the tasks of operating the equipment or monitoring the process, will your new work occur in a cockpit, where you operate and diagnose whole arrays of machines and processes, just like the oil drillers we discussed earlier?

Step Three: Identify Automation Options

Now, consider the three types of automation (in chapter 3): RPA, cognitive automation, and social robotics. Determining which types of automation will apply to your work will help you understand the likely speed and consequences of automation for you. Remember that in most real situations, several types of automation will converge.

RPA applies best if your job has mental tasks with one or more of the three Rs: repetition, redundancy, and risk. RPA needs no programmed intelligence, so it is typically the cheapest, quickest, and easiest to apply. If your tasks are amenable to RPA, you can probably assume they will not be done by you or other humans.

Human workers and their union or other representatives often resist RPA, fearing that they will be displaced. That is a very real possibility. However, the long-term solution will seldom prevent the new technology. Rather, the optimum and most socially beneficial answer is likely to emerge as workers and leaders openly and regularly discuss the implications, and reinvent the jobs before humans are displaced. A recent *Chicago Tribune* article described a Wisconsin factory where robots don't replace humans, but they fill in the gaps caused by worker shortages.[7] On an assembly line intended for twelve workers, "two were no-shows. One had just been jailed for drug possession and violating probation. Three other spots were empty because the company hadn't found anybody to do the work. That left six people on the line jumping from spot to spot, snapping parts into place and building metal containers by hand, too busy to look up." Despite increases in wages and efforts to recruit qualified and stable workers, the shortage persisted.

RPA and social robots provided a win-win solution: "In earlier decades, companies would have responded to such a shortage by either giving up on expansion hopes or boosting wages until they filled their positions. But now, they had another option. Robots had become more affordable. No longer did machines require six-figure investments; they could be purchased for $30,000, or even leased at an hourly rate." Supervisors explained that the robots would not take anyone's job, but make up for jobs they couldn't fill. One worker, reliably doing the job for years, tired of the constant turnover and empty work spaces that required her to work harder. When one employee said, "They are taking somebody's job," she replied "No, they are not . . . It's not a good job for a person to have anyway."[8] Cognitive automation can do things like recognize patterns, understand language, and learn rules. Which of your tasks does these things? If the patterns are simple, the languages are well known, and the rules are fixed, then cognitive automation or AI may substitute for humans, or already has. The move to automation is even faster when cognitive automation can be combined with sensors, such as the cameras equipped with AI that know they are photographing.[9]

On the other hand, the more complex or obscure the patterns, languages, and rules of your task, the longer cognitive automation will take to master them. Indeed, in the interim between fully human and fully automated, the new human work is training the cognitive automation. Human riders train the AI that supports self-driving cars. Every time the human corrects the AI, the program learns. Perhaps your work will evolve from *doing* the task to *training the AI* to learn the task. Uber built a fake city in Pittsburgh, called Almono, populated entirely by self-driving cars. In the fake city, automation designers can simulate things that they can't test in the real world, such as mannequins darting out front of vehicles. In addition to vehicle testing, an important goal is to train the human drivers who will eventually accompany the vehicles on actual roads: "The program that trains vehicle operators is rigorous. It takes three weeks to complete and requires trainees to pass multiple written assessments and road tests."[10]

Of course, cognitive automation may soon teach itself. In October 2017, the AlphaGo Zero AI system achieved what some have called a "singularity" in which it learned to play the board game Go with no human interaction. Designers simply programmed the game with the rules and objectives, and then instructed it to play by itself, over and over, and very quickly. An earlier AlphaGo system had required programming with the past games human experts had played. That first system defeated a world champion player in March 2016. The new system, playing only by itself, beat the original AlphaGo.[11] So, be alert to continuing evolution. Perhaps, in the near term, you will teach automation, like the original AlphaGo, but eventually automation like AlphaGo Zero will learn on its own, so you will need to reinvent again.

Social robotics involves robots that move and physically interact with humans. The robot Baxter is a cobot that literally works next to humans to do things like loading, machine tending, and materials handling. *Fast Company* described cobots that greet guests at the Henn na Hotel (translation "strange hotel"), and make coffee and clean rooms at the YOTEL New York; "Botlrs" at Starwood Hotels that use the elevator without assistance to deliver amenities to guests; an OSHbot at Lowe's stores that locates items for customers, and an AI security guard called Bob.[12] At the MIT Sloan School of Management, staff who are working from home can attend meetings, as a robot.

A *Financial Times* article points out the need for workers to constantly analyze how their jobs will be reinvented.[13] It describes a facility with a traditional human-worker assembly line on one side and a cobot-enhanced line on the other:

> On one side, the light is dim and workers stand at long assembly lines repeating the same task over and over. On the other, a fleet of low-lying robotic trucks scoot around the shop floor, restocking restyled workstations. In these small cells, a single employee helped by a robotic workbench assembles a virtually complete drive system that will be used to power the production of everything from cars to cola. Elsewhere, a robotic arm called Carmen helps workers load machines or pick components out of bins. Here, the light is brighter, and the workers say they are happier. "Everything is just where I need it. I don't have to lift up the heavy parts," says Jürgen Heidemann, who has worked there for 40 years, since he was 18. "This is more satisfying because I am making the whole system. I only did one part of the process in the old line." Stéphane Maillard, a 13-year veteran aircraft assembler said the robot has not replaced his job. "It has changed the way of working . . . Before it was very manual. Now it is more about piloting the robot. 100 per cent of our operators would never go back."[14]

Such low-cost cobots now allow small local factories that could not compete before to operate competitively and retain jobs for humans. Yet, the article also noted that many companies refused the reporter's request to see their cobots in action, perhaps fearing adverse publicity. Tony Burke, assistant general secretary of the Unite union, noted that "job losses could be horrendous in some areas . . . but the problem is that nobody really knows."[15] The solution is not to deny the effects or to hide from the press. Progress will require difficult, candid, and open conversations about the specific effects of automation at the work-task level. The framework in this book can help define and guide those conversations, to move beyond simplistic questions focused on job losses to a more nuanced conversation about the optimal work-human combinations, with society, workers, unions, companies, and communities defining "optimal."

Step Four: Optimize Work

The first three steps, applied to your work and the work of your colleagues, prepares you to have the right conversation about how the work will evolve and how to prepare for it. The key is to distinguish and integrate the elements of our framework: work tasks, ROIP, automation type, and automation role

John Donahoe, the CEO of ServiceNow, put it well: "There's this assumption that it's going to be people or robots, all or nothing. My experience is that it doesn't operate that way. It's automating part of the job, but not the full job. Repetitive, manual work—no one who's doing it is really enjoying it. Technology replaces and creates. It replaces manual work and creates new opportunities—new tasks, if you will. And productivity creates growth, which creates new kinds of work. It is a virtuous cycle. It's so easy to talk about it in binary terms. I just don't think that's the reality."[16]

Step Five: Navigate the Organization

Figuring out how your tasks will evolve and how to reinvent your job is not the end of the conversation. Your work will exist in a larger organization that is also being reinvented. That organization will not be confined by the traditional boundaries, with employees inside doing the work and others outside waiting to join. Rather, the future organization will be a hub of work arrangements like the ones we discussed in the previous chapter: free agents, talent platforms, volunteers, alliances, outsourcing, robotics, and AI.

You must navigate this organization by finding the optimum way that your reinvented job fits and connects with those who are your clients and collaborators. Sometimes you will be a traditional employee, sometimes a freelancer on a platform, and sometimes you'll take a tour of duty with an alliance partner, and so on.

You will face questions like these:

- What part of your future work should be kept in a traditional job?

- What proportion of future work should you pursue as a side gig through a work platform?

- What tasks will be substituted versus augmented by automation?

- How should you get the experiences and skills that will support your evolution?

- In what tasks and jobs should you become an expert, available to train AI or humans?

Principles Guiding the Development and Use of AI

The necessary conversation about AI is already underway. In October 2017, the Information Technology Industry Council (ITI), a global trade group for technology companies like Apple and Google, released principles designed to guide the development and use of AI, and the responsibilities of key parties such as companies, governments, and public-private partnerships. An excerpt from an executive summary follows:[a]

"Artificial Intelligence (AI) is a suite of technologies capable of learning, reasoning, adapting, and performing tasks in ways inspired by the human mind. With access to data and the computational power and human ingenuity required to extract increasing value from it, researchers are building intelligent software and machines to enhance human productivity and empower people everywhere. Startups, medium-sized companies, and larger technology companies have all developed AI systems to help solve some of society's most pressing problems, from medical diagnosis to education to economic productivity and empowerment.

While it is impossible to predict the full transformational nature of AI, like technological evolutions before it, we expect the potential implications to be vast. To ensure that AI can deliver its greatest positive potential, the Information Technology Industry Council (ITI)—the global voice of the tech sector—takes industry's responsibility seriously to be a catalyst for preparing for an AI world. In our Policy Principles, we outline specific areas where industry, governments, and others can collaborate, as well as specific opportunities for public-private partnership. To advance these principles, which we expect will evolve alongside AI technology, we acknowledge the following:

Industry's Responsibility in Promoting Responsible Development and Use: We recognize our responsibility to integrate principles into the design of AI technologies, beyond compliance with existing laws. While the potential benefits to people and society are amazing, AI researchers, subject matter experts, and stakeholders should continue to spend a great deal of time working to ensure the responsible design and deployment of AI systems, including addressing safety and controllability mechanisms, use of robust and representative data, enabling greater interpretability and recognizing that solutions must be tailored to the unique risks presented by the specific context in which a particular system operates.

The Opportunity for Governments to Invest in and Enable the AI Ecosystem: We encourage robust support for research and development (R&D) to foster innovation through incentives and funding. As the primary source of funding for long-term, high-risk research initiatives, we support governments' investment in research fields specific or highly relevant to AI, including: cyber-defense, data analytics, detection of fraudulent transactions or messages, robotics, human augmentation, natural language processing, interfaces, and visualizations. We also encourage governments to evaluate existing policy tools and use caution before adopting new laws, regulations, or taxes that may inadvertently or unnecessarily impede the responsible development and use of AI. This extends to the foundational nature of protecting source code, proprietary algorithms, and other intellectual property. Failure to do so could present a significant cyber risk.

The Opportunity for Public-Private Partnerships (PPPs): Many emerging AI technologies are designed to perform a specific task, assisting human employees and making jobs easier. Our ability to adapt to rapid technological change, however, is critical. That is why we must continue to be prepared to address the implications of AI on the existing and future workforce. By leveraging PPPs—especially between industry partners, academic institutions, and governments—we can expedite AI R&D, democratize access, prioritize diversity and inclusion, and prepare our workforce for the jobs of the future."

a. Information Technology Industry Council, "AI Policy Principles: Executive Summary," October 2017, https://www.itic.org/dotAsset/50ed66d5-404d-40bb-a8ae-9eeeef55aa76.pdf.

All the while, you must also be alert to the perpetual work upgrades that are the new reality of your work.

As a leader, you must create an environment where you collaborate with workers and other constituents to answer a parallel set of questions for your organization:

- What future work should be kept in a traditional job?

- What future work should come through gigs on a freelance platform?

- How should you give your workers the experiences and skills that will support their evolution (such as tours of duty, projects, social learning, etc.)?

- For what areas of work should you transform those that do the work to become the trainers of AI?

All the while, as a leader, you will also observe the perpetual work upgrades that are the new reality of your work. But, as a leader, you will also consider not just your work but related work and how to create the organization of the future.

The good news is that if organizations, nations, governments, and workers are empowered to work together, you are more likely to have the tools and structures you need. The new work ecosystem with constantly upgraded and reinvented jobs has the potential to empower workers, create boundless opportunities for careers and learning, and solve thorny issues of skill gaps and regional inequality. The bad news is that if workers and organizations navigate this new world in secret, without trust and transparency, then the work ecosystem risks being exploitive, opaque, and dysfunctional. That trust and transparency requires processes like "reinventing jobs" that offer a framework and language to candidly and clearly communicate about the coming evolution. (See the sidebar "Principles Guiding the Development and Use of AI.")

As a leader, you can be a role model, supporting the conversations with opportunities, a framework, and the necessary information and safety net that empowers you and your workers to collaborate to optimize work automation. You can help your workers follow these steps. You can create a safe culture in which workers can tell you when they see the possibilities for automation to replace their tasks and where augmenting tasks with automation could significantly improve productivity. If workers think that such revelations will result in their dismissal or exploitation, they won't share. If that happens, then you miss the chance for the productivity enhancements, and your workers miss the chance to evolve in advance. As challenging as the conversations will be, it is likely much more painful to make these transitions when everyone hides until forced to act.

Conclusion

As we consider the many rapid advancements in automation and work, remember that people are not powerless. The future of work is entirely up to us. Whether we use technology to substitute, augment, or create work is and should always be a conscious and informed choice.

In this book, we set out get beyond the hype to describe a concrete, actionable way you can understand and prepare for automation and its effects on work and jobs in your organization. Leaders everywhere are struggling with these hard questions. We hope that the tools provided here give you a more structured, nuanced way to anticipate the choices, make the tough decisions, and lead the reinvented jobs of the future.

Appendix

The grid in table A-1 shows the patterns in work-automation choices. The examples described in chapter 4 illustrate a subset of the rows contained in the grid. The examples reveal how the characteristics and value of work combine with the role and type of automation.

How can this grid guide you in thinking about how to optimally combine work and automation? For each work element, the characteristics of work and ROIP serve as inputs with implications for the role and type of automation. Let's say you have deconstructed a job and isolated tasks that are repetitive, independent, and mental, with a negative value ROIP. In chapter 2 we explained that negative ROIP suggests the most pivotal value lies in reducing mistakes, given the steepness of the left side of the curve. What type of automation? For work that is *mental*, the two potential types of automation are either RPA or cognitive automation, as by definition, social robotics involves *physical* interaction between humans and robots. RPA is typically suited to *substitute* for human labor when work is *repetitive and mental* and most often done *independently*, because RPA can both lower the cost of work and eliminate mistakes.

In the grid, RPA will always apply to repetitive mental work and social robotics will always apply to physical work, while AI can be used for multiple different types of work. This grid will serve as a useful rubric in helping you understand the optimal combinations of work and automation.

TABLE A-1

Work-automation combinations

CHARACTERISTICS OF WORK

	Repetitive vs. variable	Independent vs. interactive	Physical vs. mental	Return on improved performance	Role of automation	Type of automation
1	Repetitive	Independent	Physical	Negative-value RCIP	Substitution to reduce mistakes	Social robotics
2	Repetitive	Independent	Physical	Constant-value RCIP	Substitution or augmentation to reduce variance	Social robotics
3	Repetitive	Independent	Physical	Incremental-value ROIP	Substitution or augmentation to improve productivity	Social robotics
4	Repetitive	Independent	Physical	Exponential-value ROIP	Augmentation to transform performance	Social robotics
5	Repetitive	Interactive	Physical	Negative-value ROIP	Substitution to reduce mistakes	Social robotics
6	Repetitive	Interactive	Physical	Constant-value ROIP	Substitution or augmentation to reduce variance	Social robotics
7	Repetitive	Interactive	Physical	Incremental-value FOIP	Substitution or augmentation to improve productivity	Social robotics

(continued)

TABLE A-1 (continued)

Work-automation combinations

8	Repetitive	Interactive	Physical	Exponential-value ROIP	Augmentation to transform performance	Social robotics
9	Variable	Independent	Physical	Negative-value ROIP	Substitution to reduce mistakes	Social robotics
10	Variable	Independent	Physical	Constant-value ROIP	Substitution or augmentation to reduce variance	Social robotics
11	Variable	Independent	Physical	Incremental-value ROIP	Substitution or augmentation to improve productivity	Social robotics
12	Variable	Independent	Physical	Exponential-value ROIP	Augmentation to transform performance	Social robotics
13	Variable	Interactive	Physical	Negative-value ROIP	Substitution to reduce mistakes	Social robotics
14	Variable	Interactive	Physical	Constant-value ROIP	Substitution or augmentation to reduce variance	Social robotics
15	Variable	Interactive	Physical	Incremental-value ROIP	Substitution or augmentation to improve productivity	Social robotics
16	Variable	Interactive	Physical	Exponential-value ROIP	Augmentation to transform performance	Social robotics
17	Repetitive	Independent	Mental	Negative-value ROIP	Substitution to reduce mistakes	RPA

18	Repetitive	Independent	Mental	Constant-value ROIP	Substitution or augmentation to reduce variance	RPA
19	Repetitive	Independent	Mental	Incremental-value ROIP	Substitution or augmentation to improve productivity	Cognitive automation
20	Repetitive	Independent	Mental	Exponential-value ROIP	Augmentation to transform performance	Cognitive automation
21	Repetitive	Interactive	Mental	Negative-value ROIP	Substitution to reduce mistakes	RPA
22	Repetitive	Interactive	Mental	Constant-value ROIP	Substitution or augmentation to reduce variance	RPA
23	Repetitive	Interactive	Mental	Incremental-value ROIP	Substitution or augmentation to improve productivity	Cognitive automation
24	Repetitive	Interactive	Mental	Exponential-value ROIP	Augmentation to transform performance	Cognitive automation
25	Variable	Independent	Mental	Negative-value ROIP	Substitution to reduce mistakes	Cognitive automation
26	Variable	Independent	Mental	Constant-value ROIP	Substitution or augmentation to reduce variance	Cognitive automation
27	Variable	Independent	Mental	Incremental-value ROIP	Substitution or augmentation to improve productivity	Cognitive automation

(continued)

TABLE A-1   (continued)

Work-automation combinations

28	Variable	Independent	Mental	Exponential-value ROIP	Augmentation to transform performance	Cognitive automation
29	Variable	Interactive	Mental	Negative-value ROIP	Substitution to reduce mistakes	Cognitive automation
30	Variable	Interactive	Mental	Constant-value ROIP	Substitution or augmentation to reduce variance	Cognitive automation
31	Variable	Interactive	Mental	Incremental-value ROIP	Substitution or augmentation to improve productivity	Cognitive automation
32	Variable	Interactive	Mental	Exponential-value ROIP	Augmentation to transform performance	Cognitive automation

NOTES

--- --- --- --- ---

Introduction

1. Kyle Smith, "Blame the ATM!" *New York Post*, June 19, 2011, https://nypost.com/2011/06/19/blame-the-atm/.

2. "Are ATM's Stealing Jobs?" *The Economist*, June 15, 2011, https://www.economist.com/blogs/democracyinamerica/2011/06/technology-and-unemployment.

3. James Bessen, *Learning by Doing: The Real Connection between Innovation, Wages, and Wealth* (New Haven, CT: Yale University Press, 2015).

4. Tamar Jacoby, "Technology Isn't a Job Killer," *Wall Street Journal*, May 20, 2015, https://www.wsj.com/articles/technology-isnt-a-job-killer-1432161213.

5. Ethan J., "Banks Getting Rid of Tellers Are Replacing Them with Video Conferencing Mini-Banks," *VC Daily*, May 16, 2017, https://www.videoconferencingdaily.com/recent-news/banks-getting-rid-tellers-replacing-video-conferencing-mini-banks/.

6. Ibid.

Part 1

1. Michael J. Miller, "AI's Implications for Productivity, Wages and Employment," *PC Magazine*, November 20, 2017, https://www.pcmag.com/article/357490/ais-implications-for-productivity-wages-and-employment.

Chapter 1

1. S. Glucksberg, "The Influence of Strength of Drive on Functional Fixedness and Perceptual Recognition," *Journal of*

Experimental Psychology 63 (1962): 36–41; https://curiosity.com/topics/the-candle-problem-from-1945-is-a-logic-puzzle-that-requires-creative-thinking-curiosity/.

2. Bouree Lam, "Life as a Teller in the Age of the Automated Teller Machine, *The Atlantic*, August 12, 2016, https://www.theatlantic.com/business/archive/2016/08/the-teller-in-the-age-of-the-atm/495671/.

3. O*Net Resource Center, "About O*Net," https://www.onetcenter.org/overview.html.

4. Peter Evans-Greenwood, Harvey Lewis, and Jim Guszcza, "Reconstructing Work: Automation, Artificial Intelligence, and the Essential Role of Humans," *Deloitte Review*, July 2017.

5. Clifford Strauss, "Texas Oil Fields Rebound from Price Lull, But Jobs Are Left Behind," *New York Times*, February 19, 2017, https://nyti.ms/2lwUfw3.

6. Michael Hammer, "Reengineering Work: Don't Automate, Obliterate," *Harvard Business Review*, July–August 1990, 104–112.

7. Ibid.

Chapter 2

1. John W. Boudreau, *Retooling HR: Using Proven Business Tools to Make Better Decisions about Talent* (Boston: Harvard Business Review Press, 2010).

2. John W. Boudreau and Peter M. Ramstad, *Beyond HR: The New Science of Human Capital* (Boston: Harvard Business Review Press, 2007).

3. Ibid.

Chapter 3

1. George Zarkadakis, Ravin Jesuthasan, and Tracey Malcolm, "The 3 Ways Work Can Be Automated," hbr.org, October 13, 2016, https://hbr.org/2016/10/the-3-ways-work-can-be-automated?autocomplete=true.

2. Leslie Willcocks, Mary Lacity, and Andrew Craig, "Robotic Process Automation at Xchanging," Paper 15/03, The Outsourcing Unit Working Research Paper Series, June 2015.

3. David Silver et al., "Mastering the Game of Go without Human Knowledge," *Nature*, October 19, 2017, https://www.nature.com/articles/nature24270.

4. Jennifer Smith, "A Robot Can Be a Warehouse Worker's Best Friend," *Wall Street Journal*, August 3, 2017, https://www.wsj.com/articles/a-robot-can-be-a-warehouse-workers-best-friend-1501752600.

5. "How Allstate and Farmers Will Use Drones to Assess Damage from Hurricane Harvey," Reuters, August 30, 2107, https://finance.yahoo.com/news/allstate-farmers-insurance-drones-assess-114721776.html.

6. Sy Mukherjee, "Coming to an O.R. Near You," *Fortune*, November 1, 2017, 50–56.

7. Carrie Printz, "Artificial Intelligence Platform for Oncology Could Assist in Treatment Decisions," *Cancer*, March 6, 2017, https://onlinelibrary.wiley.com/doi/full/10.1002/cncr.30655.

8. "IBM Watson for Oncology Platform Shows High Degree of Concordance with Physician Recommendations," American Association for Cancer Research, press release, December 9, 2016, http://www.aacr.org/Newsroom/Pages/News-Release-Detail.aspx?ItemID=983#.WmPNp66nHIU.

9. Azad Shademan et al., "Supervised Autonomous Robotic Soft Tissue Surgery," *Science Transitional Medicine* 8, no. 337 (2016): 337, http://stm.sciencemag.org/content/8/337/337ra64.

10. Carly Szabo, "Artificial Intelligence Used to Predict Chemotherapy Resistance in Breast Cancer Patients," *Specialty Pharmacy Times*, September 24, 2015, https://www.specialtypharmacytimes.com/news/artificial-intelligence-used-to-predict-chemotherapy-resistance-in-breast-cancer-patients.

Chapter 4

1. "Robotic Part Inspection with the FANUC LR Mate 200i Robot," FANUC, http://www.fanucamerica.com/home/news-resources/case-studies/Inspection-Robot-Performs-Complete-Part-Inspection-Compass-Automation.

2. Raquel Maria Dillon, "Researchers Explore New Use for Drones: Detecting Methane Leaks," *NBC Bay Area News*, March 28, 2017, http://www.nbcbayarea.com/news/local/Researchers-Explore-New-Use-for-Drones-Detecting-Methane-Leaks-417383103.html.

3. Xavier Lhuer, "The Next Acronym You Need to Know About: RPA," *Digital McKinsey*, December 2016, https://www.mckinsey.com/business-functions/digital-mckinsey/our-insights/the-next-acronym-you-need-to-know-about-rpa.

4. Richard Feloni, "Consumer-Goods Giant Unilever Has Been Hiring Employees Using Brain Games and Artificial Intelligence,"

Business Insider, June 28, 2017, http://www.businessinsider
.com/unilever-artificial-intelligence-hiring-process-2017-6.

5. Randy Bean and Thomas H. Davenport, "How AI and Machine Learning Are Helping Drive the GE Digital Transformation," LinkedIn, June 8, 2017, https://www.linkedin.com/pulse/how-ai-machine-learning-helping-drive-ge-digital-tom-davenport.

6. Ellen Messmer, "Coca-Cola Co.'s 'Black Book' Application Squeezes the Best Out of OJ," *Network World*, May 15, 2014, https://www.networkworld.com/article/2176933/applications/coca-cola-co-s-black-book-application-squeezes-best-out-of-oj.html.

7. David Kirkpatrick, "For Stitch Fix, the AI Future Includes Jobs," *Techonomy*, October 2, 2017, http://techonomy.com/2017/10/software-plus-stylists-equal-sales-stitch-fix/.

8. Alex Voica, "How Ocado Uses Machine Learning to Improve Customer Service," Ocado Technology (blog), October 13, 2016, https://ocadotechnology.com/blog/how-ocado-uses-machine-learning-to-improve-customer-service/.

9. John Huetter, "Top U.S. Insurers Using Tractable in Photo Estimating AI Pilots," *Repairer Driven News*, October 9, 2017, http://www.repairerdrivennews.com/2017/10/09/top-u-s-insurers-using-tractable-in-photo-estimating-ai-pilots/.

10. Ted Greenwald, "Chip Makers Are Adding Brains Alongside Cameras Eyes," *Wall Street Journal*, October 14, 2017, https://www.wsj.com/articles/chip-makers-are-adding-brains-alongside-cameras-eyes-1507114801.

11. Fred Lambert, "Tesla Expands on its New Car Insurance Programs as Self-Driving Technology Improves," *electrek,* February 23, 2017, https://electrek.co/2017/02/23/tesla-insurance-program-self-driving-technology/.

12. Willis Towers Watson, "Willis Towers Watson and Roost to Establish Home Telematics Consortium of U.S. Carriers," press release, May 31, 2017, https://www.willistowerswatson.com/en/press/2017/05/willis-towers-watson-roost-establish-home-telematics-consortium.

Chapter 5

1. Menno van Doorn, Sander Duivestein, and Peter Smith, "The Unorganization: Design to Disrupt," September 5, 2017, http://labs.sogeti.com/downloads.

2. This section is based on Zhang Ruimin, "Leading to Become Obsolete," *MIT Sloan Management Review*, June 19, 2017.

3. Jay Galbraith, "The Star Model," http://www.jaygalbraith
.com/images/pdfs/StarModel.pdf.

Chapter 6

1. Kevin Kelly, *The Inevitable: Understanding the 12 Technological Forces That Will Shape Our Future* (New York: Viking Press, 2016).

2. Yaarit Silverstone, Himanshu Tambe, and Susan M. Cantrell, *HR Drives the Agile Organization* (New York: Accenture, 2015).

3. Danielle D'Angelo, "Despite Hype, Few Workers Believe Artificial Intelligence Will Threaten Their Jobs," Genpact press release, November 14, 2017, http://www.genpact.com/about-us/media/press-releases/2017-few-workers-believe-artificial-intelligence-ai-will-threaten-their-jobs.

4. W. F. Cascio, J. W. Boudreau, and A. H. Church, "Maximizing Talent Readiness for an Uncertain Future," in *A Research Agenda for Human Resource Management—HR Strategy, Structure, and Architecture,* ed. C. Cooper and P. Sparrow (London: Edward Elgar Publishers, 2017).

5. Alvin Toffler, *Future Shock* (New York: Random House, 1970).

6. J. W. Boudreau, "HR at the Tipping Point: The Paradoxical Future of Our Profession," *People + Strategy* 38, no. 4 (2016): 46–54.

7. World Economic Forum, "The Future of Jobs: Employment, Skills and Workforce Strategy for the Fourth Industrial Revolution," January 2016, http://www3.weforum.org/docs/WEF_FOJ_Executive_Summary_Jobs.pdf.

8. Reid Hoffman, Ben Casnocha, and Chris Yeh, "Tours of Duty: The New Employer-Employee Contract," *Harvard Business Review,* June 2013, 48–58.

9. Oxford Economics, "Global Talent 2021: How the New Geography of Talent Will Transform Human Resource Strategies," 2012, https://www.oxfordeconomics.com/Media/Default/Thought%20Leadership/global-talent-2021.pdf.

10. "Kellogg CEO Says Closing Oldest Battle Creek Plant Key to Firms' Survival," *Lubbock Avalanche Journal,* September 5, 1999, http://lubbockonline.com/stories/090599/bus_090599120.shtml#.WmOt3ainHqh.

11. John Boudreau, "Leaders, You Can't Achieve Agility in the Workplace Without Transparency," ReWork, October 25, 2017, https://www.cornerstoneondemand.com/rework/leaders-you-cant-achieve-agility-workplace-without-transparency.

Chapter 7

1. Michael Grothaus, "Bet You Didn't See This Coming: 10 Jobs That Will Be Replaced by Robots," *Fast Company*, January 19, 2017, https://www.fastcompany.com/3067279/you-didnt-see-this-coming-10-jobs-that-will-be-replaced-by-robots.

2. James T. Austin, Gail O. Mellow, Mitch Rosin, and Marlene Seltzer, "Portable, Stackable Credentials: A New Education Model for Industry-Specific Career Pathways," McGraw-Hill Research Foundation, November 28, 2012, http://www.jff.org/sites/default/files/publications/materials/Portable Stackable Credentials.pdf.

3. Aditya Chadrabortty, "A Basic Income for Everyone? Yes, Finland Shows It Really Can Work," *The Guardian*, October 31, 2017, https://www.theguardian.com/commentisfree/2017/oct/31/finland-universal-basic-income.

4. SkillsFuture Mid-Career Enhanced Subsidy, http://www.skillsfuture.sg/enhancedsubsidy#howdoesitwork.

5. Emily Price, "Bill Gates' Plan to Tax Robots Could Become a Reality in San Francisco," *Fortune*, September 5, 2017, http://fortune.com/2017/09/05/san-francisco-robot-tax/.

6. "Artificial Intelligence Will Create New Kinds of Work," *The Economist*, August 26, 2017, https://www.economist.com/news/business/21727093-humans-will-supply-digital-services-complement-ai-artificial-intelligence-will-create-new.

7. Chico Harlan, "Rise of the Machines: At a Wisconsin Factory, Workers Warily Welcome Robots," *Chicago Tribune*, August 5, 2017.

8. Ibid.

9. Ted Greenwald, "Chip Makers Are Adding 'Brains' Alongside Cameras' Eyes," *Wall Street Journal,* October 4, 2017, https://www.wsj.com/articles/chip-makers-are-adding-brains-alongside-cameras-eyes-1507114801.

10. Danielle Muoio, "Uber Built a Fake City in Pittsburgh with Roaming Mannequins to Test its Self-Driving Cars," *Business Insider*, October 18, 2017, https://amp-businessinsider-com.cdn.ampproject.org/c/s/amp.businessinsider.com/ubers-fake-city-pittsburgh-self-driving-cars-2017-10.

11. Satinder Singh, "Learning to Play Go from Scratch," *Nature News & Views*, October 19, 2017, https://www.nature.com/articles/550336a.

12. Vivian Giang, "Robots Might Take Your Job, But Here's Why You Shouldn't Worry," *Fast Company*, July 28, 2015, https://www.fastcompany.com/3049079/robots-might-take-your-job-but-heres-why-you-shouldnt-worry.

13. Peggy Hollinger, "Meet the Cobots: Humans and Robots Together on the Factory Floor," *Financial Times*, May 4, 2016, https://www.ft.com/content/6d5d609e-02e2-11e6-af 1d-c47326021344?mhq5j=e7.

14. Ibid.

15. Andrew Nusca, "Humans vs. Robots: How to Thrive in an Automated Workplace," *Fortune*, June 30, 2017, http://fortune .com/2017/06/30/humans-robots-job-automation-workplace/.

16. Ibid.

INDEX

Abbeel, Pieter, 78
Accenture, 143
accountability, 134
accountants, 27–28
agile thinking, 154–155
agile work, 170–173
agility, 143–144, 167–168
Alibaba, 135
alliances, 146, 183
Allstate, 76
AlphaBeta Analysis, 25–26
AlphaGo Zero, 71–72, 180
Amazon, 135
Amazon Go, 71
Android programming, 156–157, 158–159
Anglo American, 33
anthropoid robots, 76
artificial intelligence (AI), 29–30
  in cognitive automation, 71, 72–73
  convergence and, 66–68
  definition of, 184
  flight attendants and, 54
  insurance claims and, 115
  job options and, 146
  in oncology patient care, 82–83, 84
  principles guiding development and use of, 184–186
  in recruiting, 100–102

assembly-line jobs, 73–76, 128–129
ATMs (automated teller machines), 2–4, 17
  optimizing bank work automation and, 57–66
  payoff of, 43–46
  work optimization and, 88–89
authority, 134
automation
  challenges in, 42
  combining human work with, 3–4
  compatibility analysis of jobs for, 8–9
  convergence in, 77–86
  effects of on work, 66
  evolution of, 78
  execution and, 41–42
  experimentation with, 12–13
  humans replaced by, 1–2, 8
  identifying opportunities for, 117
  identifying options in, 57–86, 177–182
  implementation checklist for, 117–118
  job creation and destruction by, 17
  leadership and, 143–168
  new work for humans from, 89

automation (*continued*)
  in oncology surgery, 77–86
  optimism/pessimism about, 1,
    169–170, 182
  organizational effects of,
    116–140
  organizational readiness for,
    6–8
  payoff of, 39, 41–55
  personal career tool for
    identifying, 173, 177–182
  reengineering vs.
    deconstruction and, 36
  rethinking of work and, 32–34
  ROIP and, 54–55
  tasks compatible with, 26–28
  types of, 9–10, 58, 66–77
  uneven distribution of, 66–68
  work-automation options,
    189–194
  work optimization with,
    87–118

banking
  ATMs vs. bank tellers in, 2–4
  optimizing automation in,
    57–66
  work optimization in, 88–89
Bank of America, 3
bank tellers vs. ATMs, 2–4,
    20–23
  ROIP of, 43–46
  work optimization and, 88–89
Baxter cobot, 75–76, 180
becoming, 143–144
"Berkeley Startup to Train
    Robots Like Puppets"
    (Sanders), 78
Bessen, James, 2
Beyond HR (Boudreau and
    Ramstad), 42, 49

BHP Billiton, 33
big data, 72
Black Book model, 106
Blue Prism, 70
Botlrs, 180
Boudreau, John W., 4–5, 42, 49,
    167–168
Burke, Tony, 181–182
business processes
  in cancer treatment, 136
  reengineering, 34–38, 72
  reinventing related, 109–116
  in the star model, 136

call-center agents, 28, 108–109,
    118, 134–135
cancer care, 77–86, 90–91, 98
  organizational effects of
    automation in, 136–142
capabilities, 136, 154–157
  development of, 164–166
  enabling, 154–157, 162,
    164–166
  of leaders, 166–168
career paths, 124, 149
  continuous development,
    154–155
  development and, 164–166
  personal work evolution and,
    124, 169–188
Center for Effective
    Organizations, 206
Chen, Peter, 78
claims processing, 109–116
cobots, 74–77, 180–182
Coca-Cola Company, 106
cognitive automation, 9–10,
    70–73
  bank tellers and, 60–61
  in claims processing,
    112–113

definition of, 10
employment options and,
146–147
flight attendants and, 90
identifying options for,
179–180
in oncology patient care,
81–82, 84–85
when to use, 58, 192–193
work optimization with,
100–109
collaboration, 27–28
environment of, 186–187
collaborative robotics, 74–77.
See also social robotics
in oncology surgery, 82–83
Colson, Eric, 107
communication, 27, 168
perpetual conversation about
job reinvention, 170–173
Compass Automation, 94–95
compatibility
analysis of jobs for
automation, 8–9
of social automation, 76
of tasks with automation
types, 26–28
compensation, 131–132, 136,
158–162
in the star model, 136
compliance, 98–100, 118
constant-value ROIP curve,
47–48
convergence, 66–68, 72–73
oncology surgery and,
77–86
cost cutting, 8
counterfeit goods, 150
credit analysts, 27, 99–100, 101
critical thinking, 154
CrowdFlower, 175
culture, 134

customer-care specialists,
108–109
customer relationship
management (CRM), 70
customer service
representatives, 133–134,
152–153

da Vinci Xi system, 79–80, 82
decision making, 11, 132
deployment, 162–164
Deutsche Post AG, 76
development, 164–166
digital twins, 103–105
Dixon, Desiree, 22
Donahoe, John, 182
driverless vehicles, 114–115
drones, 76, 77
in claims processing, 112–113
in methane leak detection,
97–98
Duan, Rocky, 78
Duncker's candle problem, 19

The Economist, 175
education, 148, 156–157,
170–172
empowerment, 187–188
enabling skills, 154–157, 162,
164–166
enterprise resource planning
(ERP), 70
environmental harm, 95–96
execution, 41–42
experimentation, 12–13
exponential-value ROIP curve,
44, 48–49
cognitive automation and,
102–109
social robotics and, 96–98

factories, 148, 181–182
Farmers Insurance, 76
Fast Company, 169–170, 180
financial services, 62–63, 98–100, 102. *See also* banking
flexibility, 76
flight attendants, 52–55, 90
focus, in reengineering vs. deconstruction, 36
food as a service, 129–136
Ford, 148
"The Fourth Industrial Revolution Is About Empowering People, Not the Rise of Machines" (Keywell), 74–75
free agents, 146, 183
freelance workers, 146–147, 183
Freeport-McMoRan, 33
Future of Work study, 115, 145–146
Future Shock (Toffler), 148

Galbraith, Jay, 135–136
Gates, Bill, 172
GE, 102–105
General Motors, 148
Genpact, 144
Global Future of Work Survey, 6–7
Global Talent 2021, 154–155
Google
  AI tools, 108
  DeepMind, 71–72
  Glass, 54
governance, 118, 172
government, 170, 187–188
Gutierrez, Carlos, 167–168

Haier, 129–136, 142, 144–145
Hammer, Michael, 35–38

Henn na Hotel, 180
HolacracyOne, 128
Hsieh, Tony, 127–128
human capital perspective, 5
"human in the loop," 175
human resources, 136
  in the star model, 136
human resources consultants, 27
Hurricane Harvey, 76–77

IBM, 171
IBM Watson, 109
IBM Watson for Oncology (WFO), 81–82
implementation costs, 67
incremental-value ROIP curve, 44, 48
  social robotics and, 96
  work optimization for, 98–100
independent vs. interactive tasks, 9, 27–28
  automation optimization and, 59–61
  bank tellers, 22–23
  social robotics for, 96–98
The Inevitable (Kelly), 143
information sharing, 11, 134
Information Technology Industry Council (ITI), 184–186
innovation, 128–129, 185–186
inspection robots, 94–95
Institute for International Finance, 98
insurance industry, 109–116
integration or robots, 76
interactive tasks. *See* independent vs. interactive tasks
Internet of Things (IoT), 115, 134, 135

Intuitive Surgical, 79–80, 82
iPhone, 143–144
iRig, 96

Java programming, 156
JD, 150–153
Jenga, 125, 126
Jesuthasan, Ravin, 4
Jet Propulsion Laboratory, 97
job architectures, 162–164
job-automation matrix, 151–153
job deconstruction, 5, 6–7,
    19–39
  ATMs/bank tellers and, 20–23
  history of, 34–39
  leaders in, 144–145
  oil drillers, 28–34
  personal career tool for, 173,
    174–175
  pilots and flight attendants,
    52–55
  recruiters, 101–102
  reengineering vs., 34–38
  into work elements, 24–26
  work optimization and, 88–89,
    92–94
job reinvention, 10–11
  bank tellers, 3–4
  deconstruction in, 5, 6–7
  employee skills and, 136,
    154–157
  enabling skills and, 154–157,
    162, 164–166
  leadership in, 144, 145–166
  need for continuous, 181
  oil drillers, 28–34
  in oncology surgery, 77–86,
    136–142
  to optimize work automation,
    58–66
  options in, 146

organizational effects of,
    125–127
  payoff differences and, 88–89
  perpetual conversation about,
    170–173
  personal, 124, 169–188
  personal career tool for,
    172–188
  in process reengineering and
    automation optimization, 39
  of related jobs, 77–86, 109–116
  reskilling and, 86, 164–166
  road map for, 4–8
  work optimization and, 116
  work readiness and, 155–157
job replacement, 169–170, 182
  ATMs and bank tellers, 2–4,
    62–63
  evaluating jobs for, 8–9
  optimism/pessimism about, 1–2
job skills, 136, 144, 154–157

Kates, Amy, 136
Kellogg Company, 167–168
Kelly, Kevin, 143
Kensho Technologies, 102
Kessler, Greg, 136
Keywell, Brad, 74–75

Lawler, Edward, 136
leaders and leadership, 123,
    143–168
  capabilities in, 143–144
  checklist for automation
    implementation for, 117–118
  collaboration with workers
    in, 146
  deployment and, 162–164
  development and succession
    of, 145

leaders and leadership
(*continued*)
dispersed, 144–145
employee skills and, 155–157
environment of collaboration
and, 186–187
framework for job reinvention
for, 5–8
future capabilities for, 3–4
mindset in, 147, 148–155
reward systems and, 158–162
skills for, in the future,
166–168
transformative changes
redefining, 147
trust and, 11–12
in work optimization, 187–188
work reinvention and, 11, 144,
145–166
Lead the Work (Boudreau,
Jesuthasan, and Creelman),
136, 171
learning
educational institutions and,
171–172
evolution of automation and,
78
mindset of reiterative, 147,
148–155
work readiness and, 156–157
Learning by Doing (Bessen), 2
LinkedIn, 101, 158, 163
logistics, 153
Lowe's, 180
LR Mate 200iC robot, 94–95
Lynda.com, 156–157

machine learning, 71, 104–105,
180
maintenance optimization,
102–105

Massachusetts Institute of
Technology (MIT), 17, 123,
180
McDonald's, 49–52
McKinsey & Company, 99
Memorial Sloan Kettering
Cancer Center, 81
mental tasks. *See* physical vs.
mental tasks
methane leaks, 96–98
metrics, 136, 138
Metso, 33
mindset, 147, 148–155, 167–168
mistake reduction, 44, 46–47,
99–100
bank tellers and, 88–89
Mohrman, Susan, 136
Mutual Benefit Life (MBL), 36–38

NASA, 97
natural resources extraction,
28–34, 96, 159–162
Navy Federal Credit Union, 22
negative-value ROIP curve,
46–49
for pilots, 90
social robotics and, 94–96
work optimization for, 98–100
Nabors Industries, 96

Oakley, Eloy Ortiz, 171–172
Obama, Barack, 2
Ocado Group, 108–109
oil drillers, 28–34, 159–162
oncology surgery, 77–86, 90–91,
98, 136–142
O*Net, 24–26
organizational perspective, 5, 116
outside-in vs. inside-out
approach with, 127–142

organizational structure, 11,
    116, 123
  cancer care automation and,
    136–142
  at Haier, 129–136
  leadership and, 144–145
  personal career tool for
    navigating, 173, 183–188
  reinvention of, 142
  star model of, 135–137
OSHbot, 180
outsourcing, 146, 183
oversight, 118
Oxford Economics, 154–155

Pacific Gas & Electric (PG&G),
    97–98
pattern recognition, 70–71
performance
  assessing ROIP for, 176–177
  McDonald's vs. Starbucks,
    49–52
  payoffs of changes in, 88–89
  pilots and flight attendants,
    52–55
  ROIP, 43–55
  strategic value and, 5
  value of, 9
personal devices, banking
    and, 3
personal work evolution,
    169–188
physical vs. mental tasks, 9, 28
  automation optimization and,
    59–61
  automation types for, 189–194
  bank tellers, 22–23
pilots, 52–55
Pioneer Natural Resources, 34
pipe running, 96
Politico, 170

power structures, 134–135, 148
predictability, 148
procurement planners, 105–107
product development, 72,
    105–107, 128–129
Project Z, 151
public-private partnerships, 186
Python programming, 156

Quiet Logistics, 76

Ramstad, Peter, 42
recruiting, 100–102
redeployability, 76
redundant steps, RPA and,
    68–69
reengineering, 34–38, 72
"Reengineering Work: Don't
    Automate, Obliterate"
    (Hammer), 35–38
regulatory risk, 70
relationship skills, 154–155
repetitive vs. variable tasks, 8,
    26–27
  automation optimization and,
    59–61, 190–194
  bank tellers, 22–23, 88–89
  RPA and, 68
  social robotics for, 94–96,
    96–98
research and development
    (R&D), 185–186
reskilling, 86, 164–166
return on improved
    performance (ROIP),
    43–46
  automation optimization and,
    59–61
  automation types and, 67
  curves in, 46–49

return on improved
performance (ROIP)
(*continued*)
McDonald's vs. Starbucks,
49–52
personal career tool for, 173,
176–177
pilots and flight attendants,
52–55
reward systems linked to,
158–162
social robotics and, 94–96
work-automation grid on,
190–194
work optimization and, 88–89,
90–94
return on investment (ROI), 118
reward systems, 131–132, 136,
158–162
Rio Tinto, 96
risk management, 44, 118
in insurance and smart-home
devices, 115
RPA and, 69
robotic process automation
(RPA), 8, 9–10, 66–70
in claims processing, 112–113
definition of, 10
employment options and,
146–147
identifying options for,
178–179
in oncology patient care, 80
scalability and, 100
tasks suitable for, 68–70
when to use, 58, 189–194
work optimization with,
98–100
"Robotic Process Automation
101" (Surdak), 68–70
robotics, 146
robotic surgery, 79–80, 82

robots, tax on, 172
Roost, 115

safety, 76
salary.com, 158
salespeople, 118
Sanders, Robert, 78
scalability, 100
self-organization, 127–128
sensors
in cognitive automation, 71
cognitive automation and, 179
oil drillers and, 29–30
serial entrepreneurship,
130–132
ServiceNow, 182
Simply Orange, 106
singularity, 180
Sloan Management Review, 130
smart-home devices, 115
Smart Tissue Autonomous
Robot (STAR), 82–83, 98
social networks, 134
social robotics, 9–10, 73–77
definition of, 10
identifying options for,
178–179, 180–182
for pilots, 55, 90
for repetitive, independent,
physical work, 94–96
when to use, 58, 190–194
SogetiLabs, 127–128
stability, 148
stackable credentials, 171–172
Starbucks, 49–52
star model of organization
design, 135–137
Starwood Hotels, 180
Stitch Fix, 106–107
strategic value, 39, 41–55
cobots and, 75–76

job performance and, 5
in reengineering vs.
    deconstruction, 36–38
of technology, 29
work performance and, 9
strategy, 5, 133
reengineering vs.
    deconstruction and, 36
in the star model, 136
stylists, 106–107
supply chain, 153
Surdak, Christopher, 68–70
swarm robots, 76–77
systemic change, 12–13

talent management, 100–102
deployment, 162–164
development, 164–166
leadership development, 145
reskilling for job changes, 6–7
skill development, 136,
    154–157
in the star model, 136
talent platforms, 146, 183
tax on robots, 172
technology
convergence in, 66–68
disruption from, 143–144
Tesla, 114–115
Toffler, Alvin, 148
trainability, 76
transparency, 168, 187–188
trial lawyers, 109
trust, 11–12, 134, 187–188

Uber, 179–180
unemployment benefits and
    insurance, 172
Unilever, 100 101
Unite, 182

universal basic income (UBI),
    172
"The Unorganization"
    (SogetiLabs), 127–128
Uptake, 74–75
Upwork, 156, 158–159, 162, 163

variable tasks. See repetitive vs.
    variable tasks
variance reduction, 44, 47–48
virtual collaboration, 142
volunteers, 146, 183

warehouse operations, 76–77,
    150
Watson for Oncology (WFO),
    81–82
Willis Towers Watson, 39, 115,
    145–146, 154–155
Winby, Stu, 136
work, evolution of, 148–154,
    174–175
worker injury, 95–96
worker shortages, 178
work optimization, 5, 10–11,
    87–116
across job groups, 24, 25
automation types and, 58–66
banking automation and,
    57–66
checklist for leaders, 117–118
of claims processing, 109–116
framework for, 91–94
job deconstruction in, 24–26
on oncology surgery, 83
personal career tool for, 173,
    182
questions guiding, 92
reinvention of job families/
    processes and, 109–116

work optimization (*continued*)
of repetitive, independent,
mental work with
incremental-value ROIP,
100–102
of repetitive, independent,
mental work with negative-
value and incremental-
value ROIP, 98–100
of repetitive, independent,
physical work with
negative-value ROIP,
94–96
of repetitive, interactive,
mental work with
exponential-value ROIP,
102–105
of repetitive, interactive,
physical work with
incremental-value ROIP, 96
of variable, independent,
mental work with
exponential-value ROIP,
105–107
of variable, independent,
physical work with
exponential-value ROIP,
96–98
of variable, interactive,
mental work with
exponential-value ROIP,
108–109
of variable, interactive,
physical work with
exponential-value ROIP, 98
work readiness, 155–157
work tasks. *See also* job
deconstruction
automation optimization and,
59–61
automation types and, 66, 67
cognitive automation and,
70–73, 72–73
deconstructing jobs into, 5,
6–7
pilots and flight attendants,
52–55
World Economic Forum, 67, 149
Worley, Christopher, 136

Xchanging, 65

YOTEL New York, 180

Zappos, 127–128, 144–145
Zhang Ruimin, 130–132, 134
Zhang Tianhao, 78

ABOUT THE AUTHORS

RAVIN JESUTHASAN is a recognized global thought leader and author on the future of work, automation, and human capital. He has written numerous research reports and articles on these topics and is the coauthor of the books *Lead the Work* (Wiley, 2015) and *Transformative HR* (Wiley, 2011). Jesuthasan is an advisor to some of the largest companies in the world and has led numerous large-scale, global restructuring and transformation engagements.

Jesuthasan is a regular participant and presenter at the World Economic Forum's annual meetings in Davos and Dalian/Tianjin and is a member of the WEF's Steering Committee on Work and Employment. He has been featured extensively by leading business media including CNN, the BBC, the *Wall Street Journal*, *BusinessWeek*, CNBC, *Fortune*, *FT*, *The Nikkei* (Japan), *Les Echos* (France), *Valor Econômico* (Brazil), *Business Times* (Malaysia), *Globe and Mail* (Canada), the *South China Morning Post*, Dubai One TV, and *The Australian*, among others. He is a frequent guest lecturer at universities around the world including Oxford University, Northwestern University, and the University of Southern California.

Jesuthasan has been recognized as one of the 25 most influential consultants in the world. He is a Managing Director of Willis Towers Watson and is based in the firm's Chicago office.

JOHN W. BOUDREAU, PHD, is Professor of Management and Organization at the University of Southern California's Marshall School of Business and Research Director at the university's Center for Effective Organizations. He is recognized worldwide for breakthrough research on the bridges between superior human capital, talent, and sustainable competitive advantage.

Boudreau's more than 200 publications include the books *Lead the Work* (Wiley, 2015), *Retooling HR* (Harvard Business Review Press, 2010), and *Beyond HR* (Harvard Business Review Press, 2007). He is featured in *Harvard Business Review*, the *Wall Street Journal*, *Fortune*, *Fast Company*, and NPR, among others. His research appears in *Management Science*, *The Academy of Management Executive*, *Journal of Applied Psychology*, *Personnel Psychology*, *Human Resource Management Review*, and *Industrial Relations*. Boudreau has received the Herbert Heneman Jr. Award for Career Achievement from the Academy of Management, the Michael R. Losey Excellence in Human Resource Research Award from the Society for Human Resource Management, and the Academy of Management's awards for Organizational Behavior New Concept and Human Resource Scholarly Achievement. He is a Fellow of the National Academy of Human Resources, the Society for Industrial and Organizational Psychology, and American Psychological Association. Dr. Boudreau advises organizations that include early-stage companies; global corporations; and government, military, and nonprofit organizations. He founded and led the Global Consortium to Reimagine HR, Employment Alternatives, Talent, and the Enterprise (CHREATE), and he serves as a Foundation Trustee of the National Academy of Human Resources and on the Transformational Investment Capacity Committee of Médecins Sans Frontières (Doctors Without Borders).

You can access a digital copy of our framework and numerous other resources to support your journey in reinventing jobs by visiting the following websites: willistowerswatson.com/reinventing-jobs and drjohnboudreau.com/speaking/reinventing-jobs-to-optimize-work-automation.